I0810508

IMAGES
of America

ALONG ROUTE 52
DELAWARE'S HISTORIC KENNETT PIKE

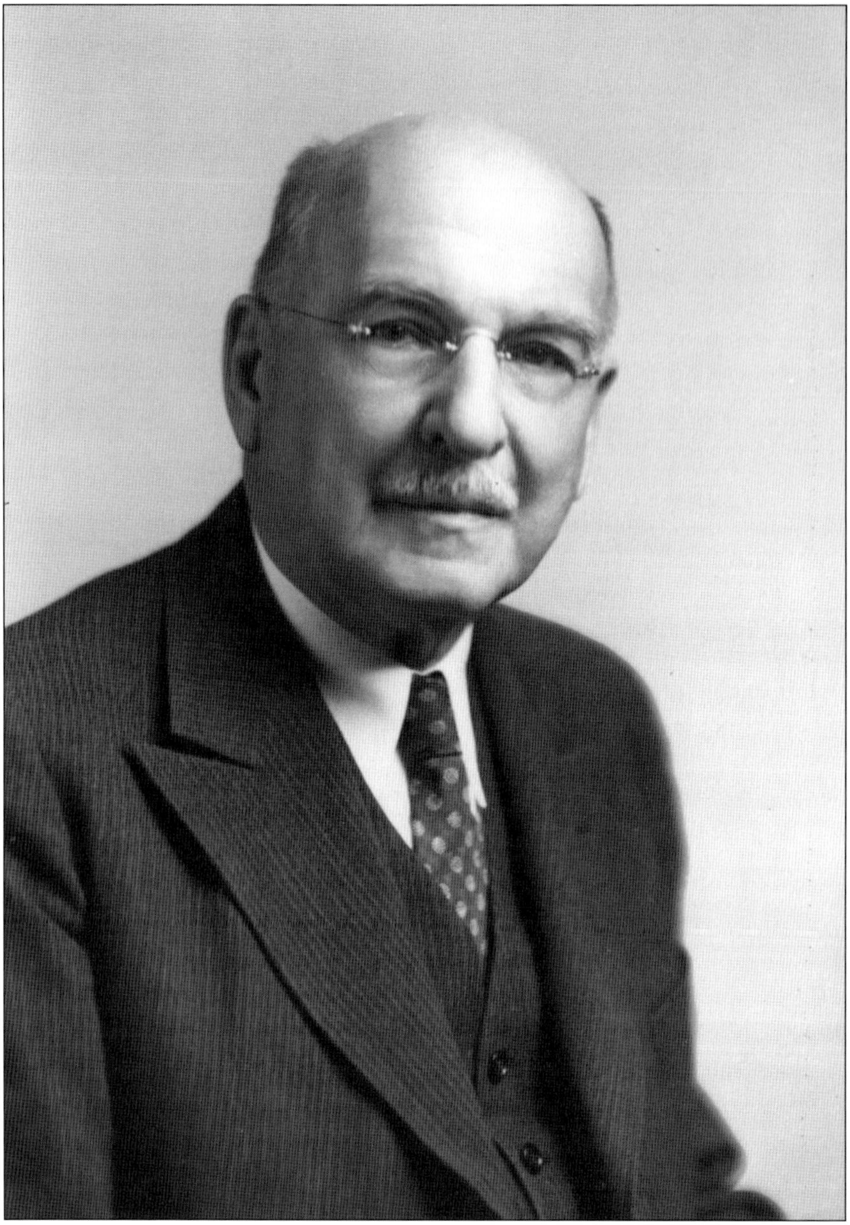

Pierre S. du Pont (1870–1954) was the benefactor of both Kennett Pike and Pennsylvania Avenue. Born into the wealthy du Pont family, Pierre was president of E.I. du Pont de Nemours & Company from 1915 through 1919. Along with his cousins T. Coleman and Alfred, Pierre helped transform the family gunpowder business into a modern corporation. His home, Longwood Gardens, is open to the public. (Courtesy of Hagley Museum and Library.)

On the Cover: Shown on the cover is Kennett Pike, just south of Brecks Lane, near Wilmington, Delaware, in 1934. To the right of the image is the present-day community of Westover Hills. This view shows the width of the pike following the construction spearheaded by Pierre S. du Pont in 1920. (Courtesy of the Delaware State Archive.)

IMAGES
of America

ALONG ROUTE 52
DELAWARE'S HISTORIC KENNETT PIKE

Andrew D. Engel

ARCADIA
PUBLISHING

ISBN 978-1-4671-2094-4

Published by Arcadia Publishing
Charleston, South Carolina

Printed in the United States of America

Library of Congress Control Number: 2013945005

For all general information, please contact Arcadia Publishing:
Telephone 843-853-2070
Fax 843-853-0044
E-mail sales@arcadiapublishing.com
For customer service and orders:
Toll-Free 1-888-313-2665

Visit us on the Internet at www.arcadiapublishing.com

To my grandparents Richard and Barbara Kane, who have inspired me more than they know and have helped shape me into who I am today

Contents

Acknowledgments

This publication would not have been possible without the assistance, guidance, and encouragement of so many individuals. First, I would like to thank my grandparents Richard and Barbara Kane. Growing up with them provided me so many opportunities that most young people do not have. Through my grandfather's construction business, I was privy to views of many local estates that few had seen. Also, my fiancée, Alexandra Schnabel, always pushes me to be the best I can, and when I am frustrated, encourages me to carry on.

I have many individuals and local historical institutions to thank for their assistance and advice. First, I am grateful to Lynn Catanese, chief curator of library collections at Hagley Museum and Library, for editing my final product and keeping me on track as well as being a great mentor and friend. At the Longwood Library, I am grateful to Judy Stevenson for her assistance with the photographic collection of Pierre S. du Pont. I also thank Maggie Lidz and Heather Clewell at Winterthur; Randy Goss at the Delaware Public Archives; and Mike Hahn at the Delaware Department of Transportation. Honorary mayor of Centreville Edward P. Frederick provided reference and photographs for the chapter on Centreville; they were a great addition to the book. George Trapnell, John Shields, Denis O'Sullivan, and Irenée du Pont Jr. also contributed photographs from their personal collections, for which I am forever grateful.

I also owe a debt of gratitude to the following people for pointing me in the direction of individuals to contact to find pictures and tales of days gone by: Phoebe Craven, Ned Frederick, Marian Frederick, Tony Lunger, Phillip Strange, Grier Wakefield, Pam Witsil, and the late Patricia Cannon.

INTRODUCTION

Created through an act of the Delaware State Legislature, the Wilmington & Kennett Turnpike was incorporated on January 21, 1811. The company was charged with laying out and providing a route connecting a growing industrial city, Wilmington, Delaware, with the southeastern suburbs of Philadelphia, Pennsylvania. In addition to serving as a primary connector to the southeastern counties in Pennsylvania, the Wilmington & Kennett Turnpike also served as a route to meet up with other turnpikes that led to the western portion of Pennsylvania. The company was a shareholder corporation comprising 800 shares, each worth 25¢.

Early after the company was organized, the state realized that the road would become increasingly important as the 19th century progressed. This is evidenced by the 100-foot right-of-way granted with the construction of the turnpike, even though only 20 feet was surfaced for use by travelers. Anyone traveling on the Kennett Pike, as it came to be known, was charged a toll based on the type of carriage (and later, vehicle) in which they were traveling, as well as amount and type of cargo. For example, 6¢ was charged for transporting a hog, 20¢ for a chair, and 12¢ for cattle. Provisions were made for those who had farms along the pike so they would not be charged when traveling from one portion of their property to another. Additionally, those traveling to a house of worship or for a funeral were not charged a fee.

With the beginning of the 20th century and the rise of the use of the automobile, travelers began to demand smoother surfaces and lower grades on which to drive. Also with the turn of the century came the rise of the modern DuPont Company, located in Wilmington. A growing number of executives from DuPont and other corporations began to make their homes and build their lavish estates along the Kennett Pike. One of those who traveled the Kennett Pike almost on a daily basis was the secretary and treasurer of DuPont, Pierre S. du Pont (1871–1954). Du Pont spent the better part of 10 years formulating a plan to modernize the Kennett Pike.

Pierre du Pont first conceived of the idea of transforming Kennett Pike into a modern highway in 1916. Through correspondence with Col. Henry A. du Pont (of Winterthur), Pierre outlined his early plans in four parts. First, he suggested that the new road be the same width from the Wilmington city limit to the state line with Pennsylvania as was Pennsylvania Avenue in Wilmington. Second, he proposed to remove all hazards—such as narrow bridges—from the road, widening culverts, and building an overhead crossing near Brecks Lane in Wilmington out of concern for greater public safety. Next, the road would be between 18 and 25 feet wide "of the most improved form." Finally, proper grading would ensure a smooth and, most importantly, safe passage of travelers. Colonel du Pont also happened to be the principal stockholder in the Wilmington & Kennett Turnpike Company, owning half of the 1,200 shares (valued at $25 apiece). Pierre then asked his cousin Henry what type of cooperation he could expect from the turnpike company if he were to pay all costs involved in the transformation of the road. Henry A. du Pont replied a few months later that he would sell all of his shares and encourage fellow shareholders to do the same. In 1917, Pierre requested a list of stockholders from his brother-in-law W.W. Laird in preparation to buy out the shareholders in their entirety, and he authorized $50,000 to purchase the shares he desired. In the meantime, work commenced and was completed on the Pennsylvania section of the road, widening and paving the approximately 1.5 miles from the Pennsylvania state line to Mendenhall. However, work on the highway proper would come to a complete halt with the entrance of the United States into World War I and would not resume again until 1919.

By April 1919, du Pont had become the virtual owner of the turnpike company. He then appointed Hugh Rodney Sharp to act as a trustee to handle the transactions, similar to the way du Pont had handled the building of Chester County Hospital. Once all shares were in his possession (at a cost of $60,000 for the shares and $10,000 for the franchise), du Pont deposited them with the Wilmington Trust Company with instructions to turn everything over to the State of Delaware in three years for the cost of $1. The first order of business before any work could be done was to form a new board of managers, which included W.W. Laird, S.D. Townsend, F.A. McHugh, R.T. Ellis, V.B. Thomas, W.M. Francis, R.W. Ellis, D.C. Aspril, L.J. Speel, and William Raskob. Du Pont then reorganized the board into a three-man committee, which included W.W. Laird as its president, S.D. Townsend as vice president/treasurer, and Charles H. TenWeeges as secretary.

Once the company was reorganized, Pierre immediately began work to widen and resurface Kennett Pike. He contracted with the engineering department of E.I. du Pont de Nemours & Company to lay out and design the highway, hiring subcontractors to do the actual construction. The principal contractor hired for the job was Delaware businessman Richard Hopkins. Hopkins and du Pont decided to start the nearly five-mile-long project approximately in the middle of the span near the Grandhomme estate, located just north of Pyles Ford Road and Winterthur. Hopkins also set up a camp for his workers at the site and later moved a camp from the Brandywine mills and placed it into operation. Once a workforce was organized, excavation began with moving a stream on the Amy E. du Pont property. By July 1919, there had been 2,700 feet of concrete poured, and du Pont and his legal team were busy securing rights-of-way in Centreville. Nearly 200 men were employed by the end of August as grading work was taking place from the Lower Brandywine Presbyterian Church to Centreville. By summer's end in September, excavation work was nearly complete on the northern stretch, and workers were moving homes in Centreville in preparation for paving. Macadam shoulders were installed beginning at the state line, and preparations were being made for an overhead bridge on the land of Mrs. Alexis I. du Pont near present-day Route 141. Du Pont finally was able to acquire land owned by his cousin William du Pont, who at the time was building Westover Hills, an upscale housing development just outside of the Wilmington City limits. With that, excavation could begin on the portion from Buck Road in Greenville to Brecks Lane. With the coming cold weather in autumn, final grading was taking place in Centreville while excavation was beginning on the Green Estate in Greenville, Delaware. The first layer of water-bound macadam was laid in early November, marking the completion of 40 percent of the work. Work came to a halt on December 13, 1919, with the coming of winter.

Work resumed again in April 1920 with dressing up the sides of the road from Greenville to the state line and placing topsoil on the disturbed sections of Winterthur. The final rights-of-way were secured by the end of spring with the exceptions of the Crossan property in Centreville and the section that would need to be acquired from the Philadelphia & Reading Railroad at Greenville. Paving was also being completed, and Kennett Pike was partially opened to traffic on June 12, 1920, and was entirely open to traffic by the end of July. Work still continued on sidewalks on the Wilmington portions.

When the widening project was finally complete, the total cost of construction for Kennett Pike totaled over $764,000. The new 24-foot-wide highway included a three-foot-wide macadam shoulder. Provisions were also made for a right-of-way to include dirt shoulders for a total width of 80 feet. Over the span of little more than a year, 200 men were employed, and they laid 4.6 miles of road, moved 86,000 cubic yards of earth and 6,300 cubic yards of rocks, relocated 31,200 feet of fence, and replaced 300 feet of hedge. Highway engineers with the state closely watched the methods of construction, including the laying the road in two longitudinal strips—instead of in a single slab—for the sake of stability. The opening of this modern highway provided travelers with the most modern standards on which to travel. At the end of the project, as planned, du Pont transferred the rights of the Kennett Pike back to the State of Delaware for $1.

Through the 1920s and 1930s, Kennett Pike became increasingly important as a modern highway but was also the front line for the growth of the American Country House movement in Delaware. Executives from the DuPont Company would continue to build their grand estates along the road from Wilmington over the state line into Pennsylvania.

The 1960s and 1970s were a time of growth on the pike, particularly on the Wilmington side, where many of the original large estates were sold off and carved up. Such was the case of St. Amour, which was the home of Mary Belin du Pont and later, her son Lammot du Pont. It was torn down in 1972 and is now the site of Tower Hill School's athletic complex. Also during this time, various improvements were made to the road, such as Kennett Pike's overhead crossing of Route 141.

Today, much of the Kennett Pike north of the Wilmington border retains its country feel. Although modern shopping centers have been established in Greenville, the pike remains largely unchanged. Centreville has undergone various road-improvement projects but remains a quaint Delaware village.

One

Wilmington

At the same time philanthropist Pierre S. du Pont (1870–1954) was planning work on Kennett Pike through Greenville and farther north into Centreville, he was also working to widen the part of the pike that lies south of Greenhill Avenue, known as Pennsylvania Avenue. Picking up where Kennett Pike unofficially ends at Greenhill Avenue, Pennsylvania Avenue carried traffic to Delaware Avenue, which continued into the center of the city, leading to the vicinity of the DuPont Building.

Work on Pennsylvania Avenue began in its planning stage in 1917 and was fully underway by 1919, commencing near Riverview Avenue by late spring of that year. Unlike other portions of the pike where du Pont only needed the permission of a handful of property owners who were primarily du Pont family members, the stretch into Wilmington involved dozens of property owners. Many property owners in the area were business associates of du Pont's, including William Coyne and Charles Patterson, who donated their land along the avenue for the widening of the road. Always a savvy businessman, du Pont did the work in stages, moving from north to south and eventually ending up close to the DuPont Building.

Pennsylvania Avenue was, and still is, home to many local landmarks. Those who live in the area have likely visited the Columbus Inn, purchased a car from Union Park (Pontiac in its early years), or even attended Ursuline, a Catholic school for young women. Some will also remember businesses such as Frienhoffer's Bakery and Diamond Ice & Coal, as well as the various service stations that dotted the avenue.

Although Pennsylvania Avenue is the road's official name in the city, the Wilmington portion of Kennett Pike, running from approximately Greenhill Avenue to the border with Greenville (just north of present-day Delaware Route 141), became the first suburban home of a number of du Pont family members. Many settled in an area just north of Tower Hill School in what was dubbed "the Neighborhood," the site of seven du Pont family homes on approximately 60 acres.

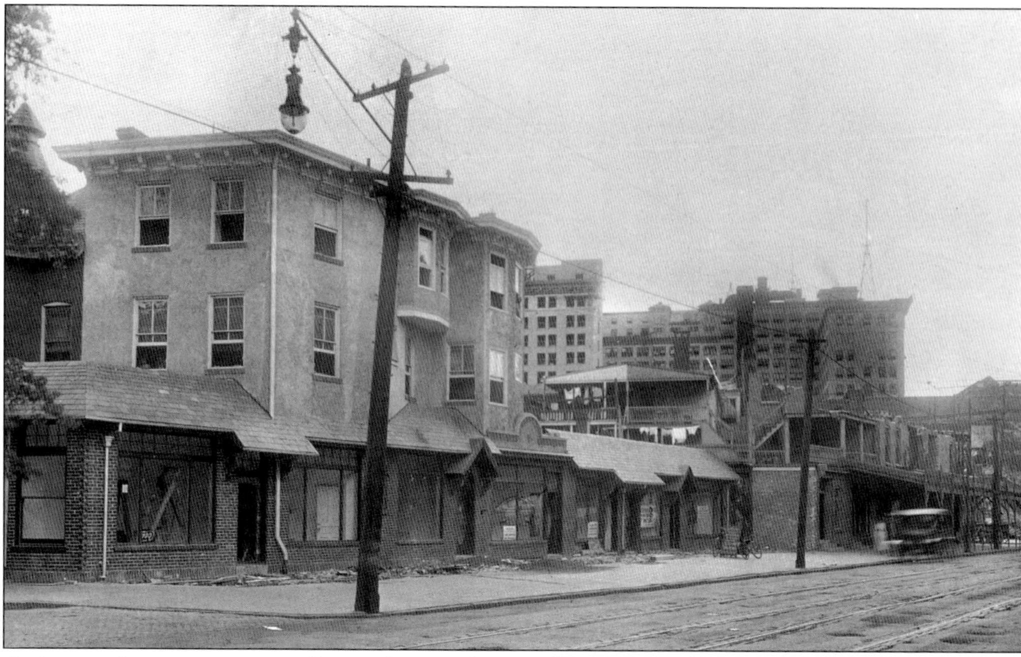

At its southernmost point, Pennsylvania Avenue becomes Delaware Avenue before reaching the headquarters of the DuPont Company. Before the stretch was widened in the 1920s, this is what one would have seen on the northeast corner of the avenue at West Street. This site is now occupied by the Brandywine Building. (Courtesy of Hagley Museum and Library.)

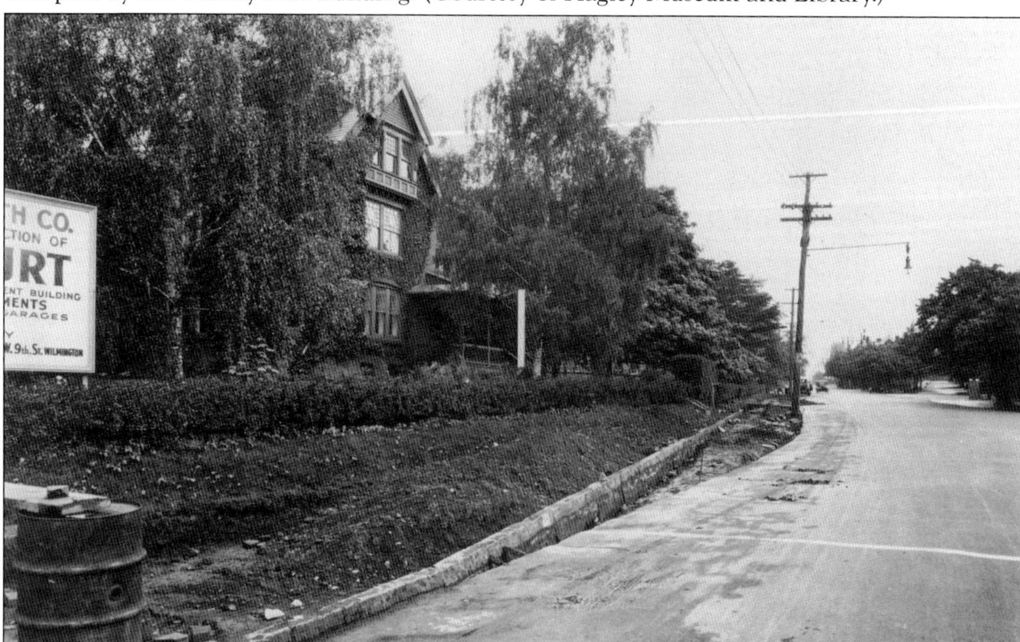

In this view looking north from downtown Wilmington, one can see the widening of Pennsylvania Avenue taking place around 1927. Located here now is a Shell gas station. Evidence suggests that, at one time, the parcel was owned by Charles Patterson of the DuPont Company. (Courtesy of Hagley Museum and Library.)

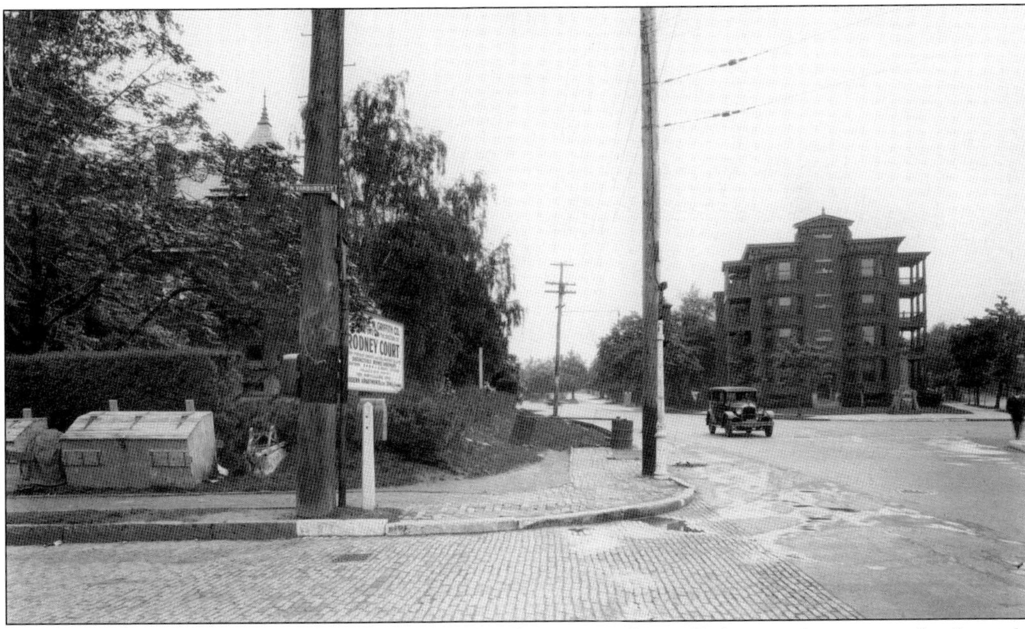

Another view looking north from the city of Wilmington shows Delaware Avenue to the far right and Van Buren Street to the left. The construction of the Rodney Court Apartments is clearly underway by the time this photograph was taken, before the widening of Pennsylvania Avenue in this vicinity in approximately 1927. The structure was the first apartment building in the state with an elevator. (Courtesy of Hagley Museum and Library.)

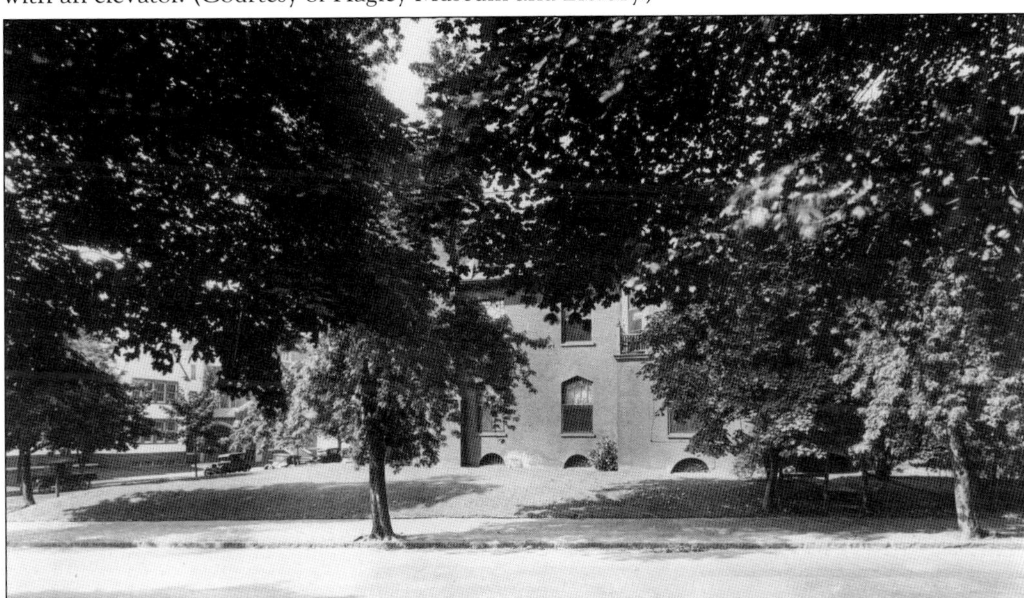

Located at the southwest corner of Franklin Street and Pennsylvania Avenue was Misses Hebbs School. Shown here before the widening of the avenue in approximately 1927, the location is now an office building. Directly across Franklin Street from this site is part of Ursuline Academy, an all-girls' Catholic school. (Courtesy of Hagley Museum and Library.)

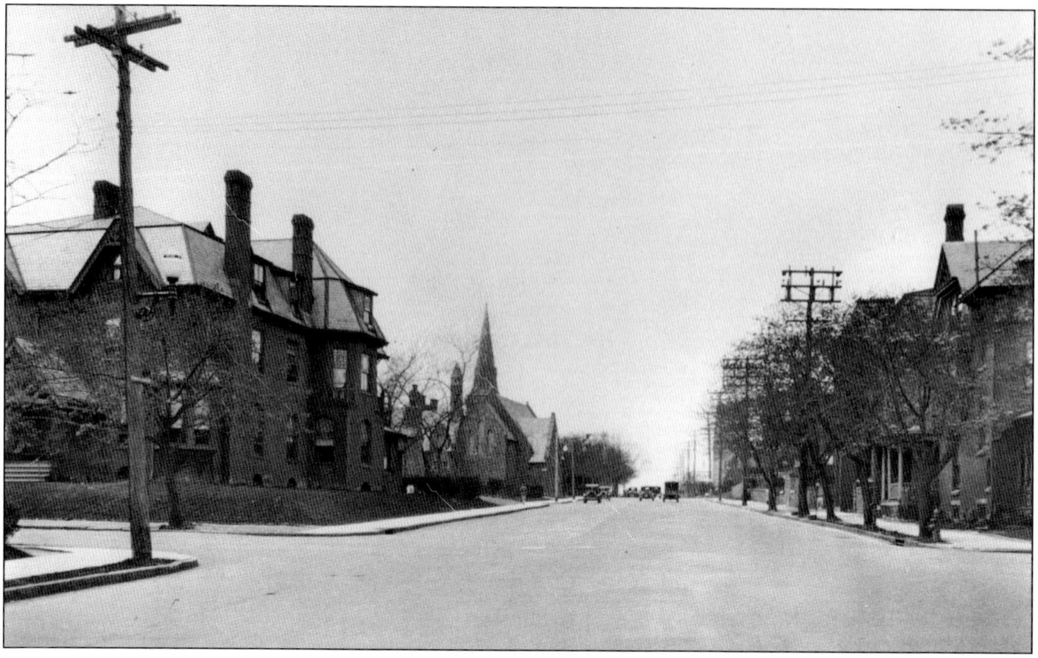

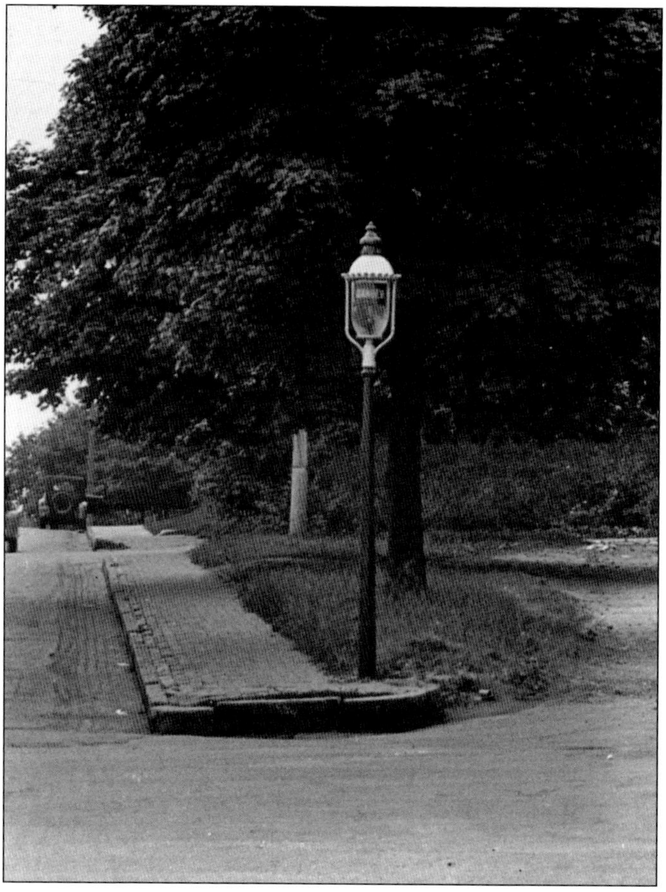

This view looking north from approximately Pennsylvania Avenue and Franklin Street shows both sides of the avenue as well as the Church of the Holy City in the distance at left. This photograph, taken around 1929, shows the avenue after construction. (Courtesy of Hagley Museum and Library.)

This view of Pennsylvania Avenue near Clayton Street shows the gas lamps that bordered the road around the year 1926. Gas-powered lamps were used before electric lines were laid for streetlights. There are only approximately 5,000 gas lamps left in operation in the United States. (Courtesy of Hagley Museum and Library.)

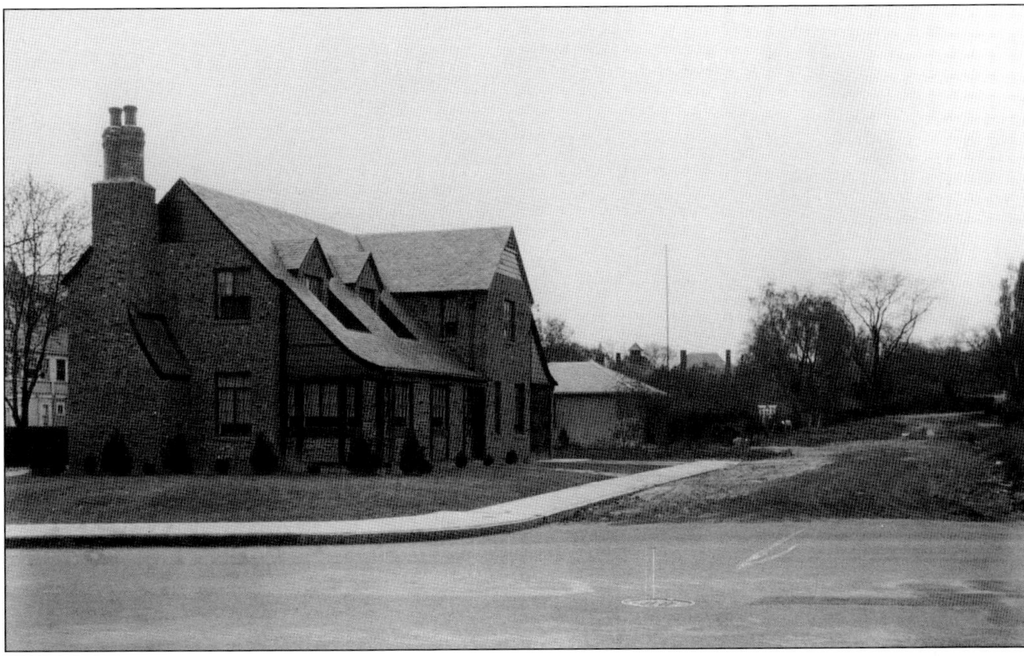

This photograph was taken just after the widening of Pennsylvania Avenue in the vicinity of Kennett Place. The dwelling shown here still stands and is a private office. Kennett Place is a cut-through to Eleventh Street near the H. Fletcher Brown Mansion and Ingleside's senior living apartments. (Courtesy of Hagley Museum and Library.)

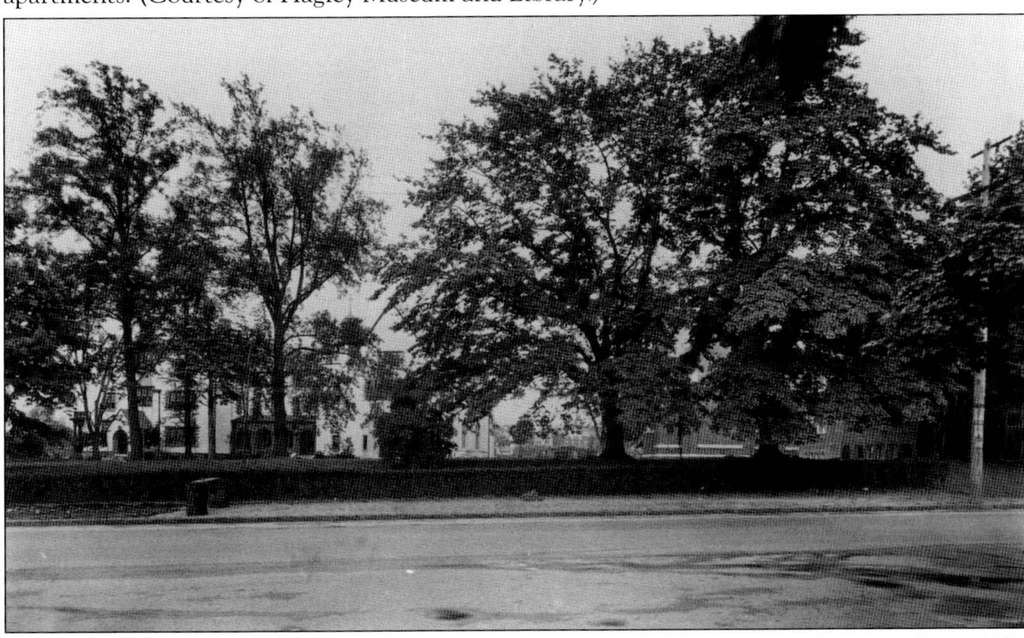

Also located in the vicinity of Franklin and Van Buren Streets is Ursuline Academy. This photograph taken in the area of Ursuline shows Pennsylvania Avenue after its widening in the late 1920s. By this time, local businessman John Raskob had donated the former Grant Estate to the academy. In 1927, there were 300 students attending the school. (Courtesy of Hagley Museum and Library.)

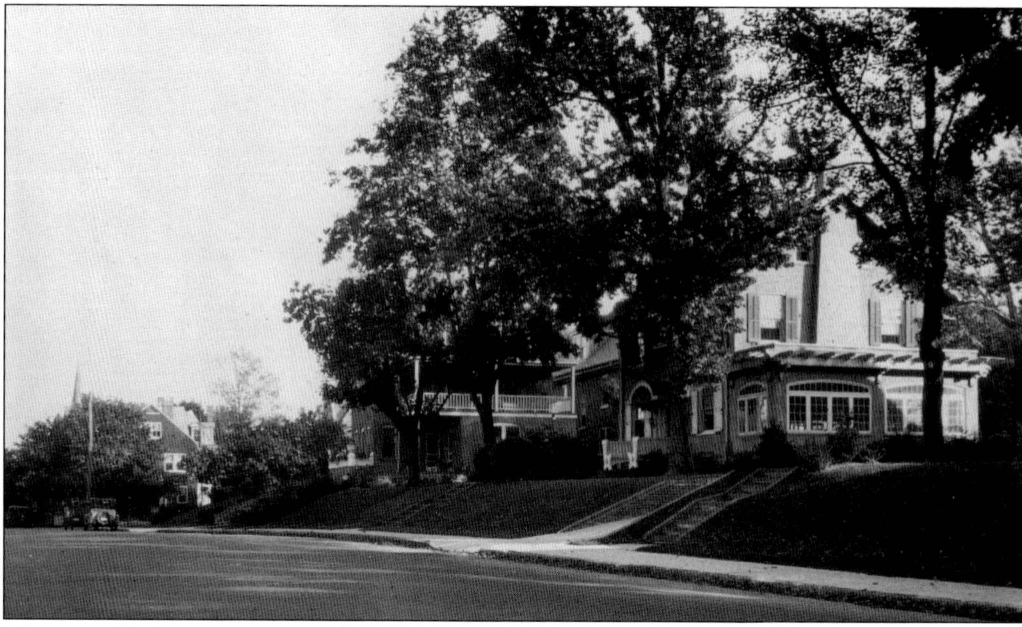

This c. 1925 photograph shows homes along Pennsylvania Avenue just north of Rodney Street (which can be seen at right) before road construction. This home at 1502 Pennsylvania Avenue was built around 1900 and still stands as a private residence. However, the front yard has been largely been taken as part of later expansions on the avenue. (Courtesy of Hagley Museum and Library.)

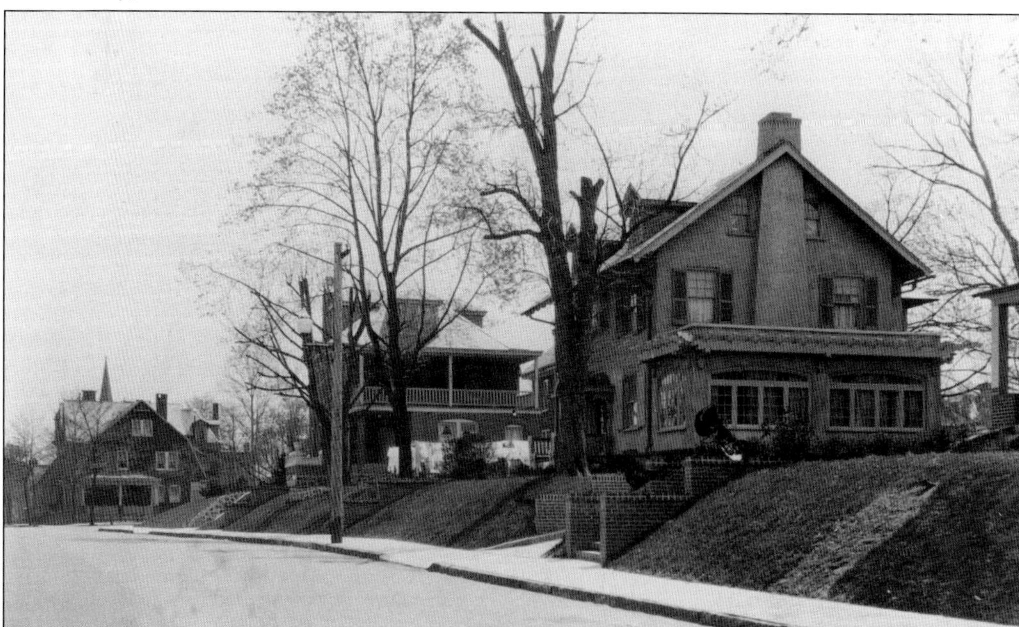

This photograph, similar to the one shown above, was taken in roughly the same spot after construction of the widened road around 1927. Viewers of the image can see how much frontage of property was lost in the name of progress. The steeple of the Church of the Holy City is visible in the background. (Courtesy of Hagley Museum and Library.)

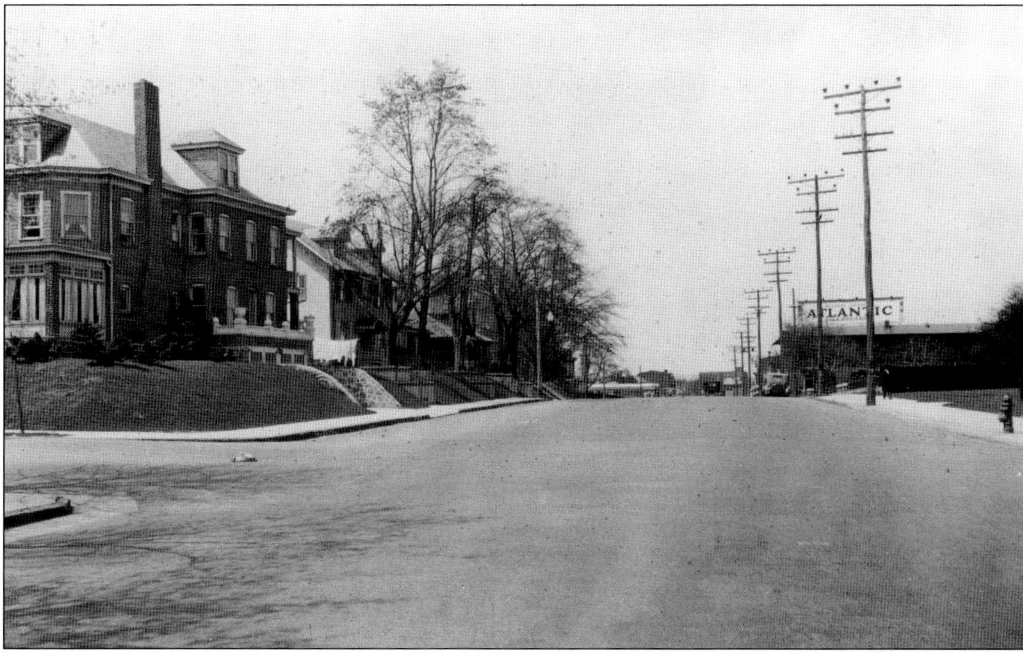

With frontage along both Pennsylvania Avenue and Rodney Street is 1109 Rodney Street. This view looking north toward Clayton Street shows Pennsylvania Avenue after construction had been completed around 1927. A sign for Atlantic Gasoline is visible in the background at the site of Kia of Wilmington. (Courtesy of Hagley Museum and Library.)

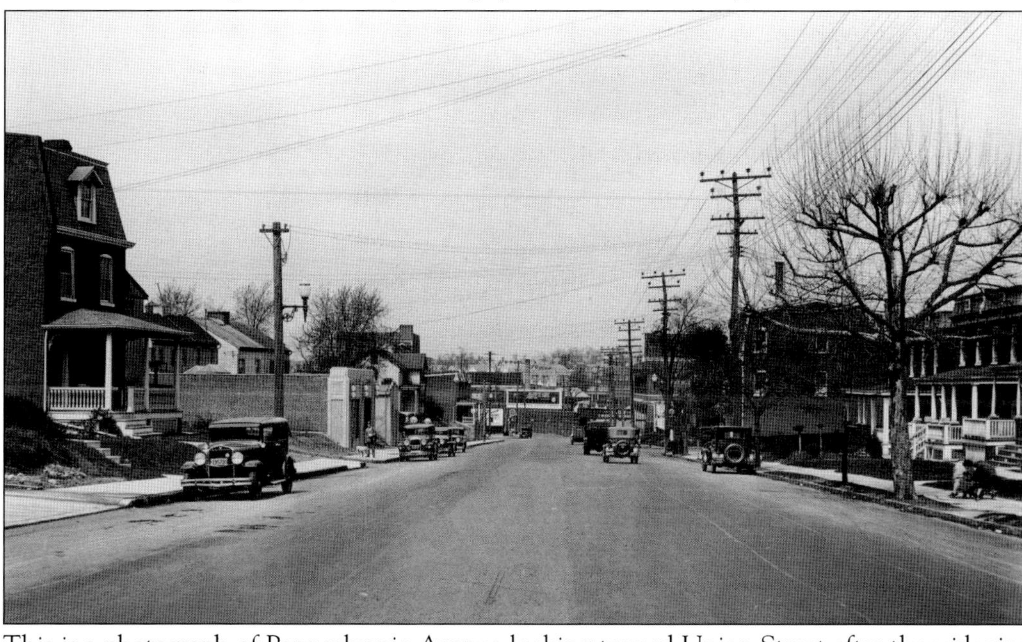

This is a photograph of Pennsylvania Avenue looking toward Union Street after the widening of the avenue around 1924. The papers of Pierre S. du Pont show that pricing and estimates for this stretch of road began as early as July 1920. The overpass of the Baltimore & Ohio (B&O) Railroad can be seen in the background. To the left is the approximate site of Delaware Cadillac today. (Courtesy of Hagley Museum and Library.)

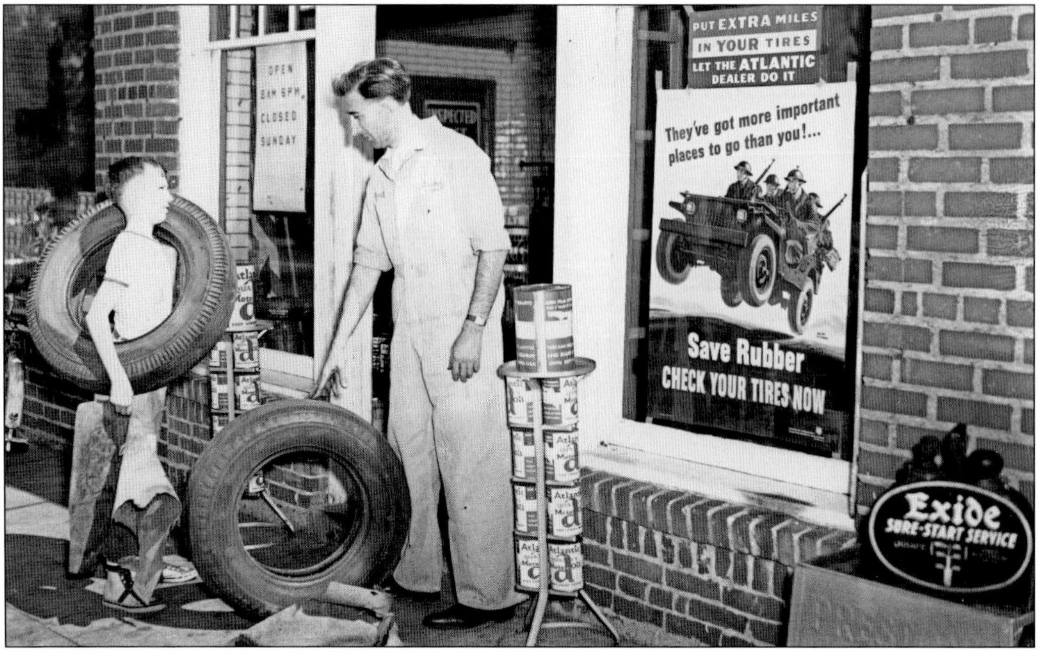

Billy Evans is captured here turning in old rubber at the Atlantic Filling Station of L.E. Wadman at 1603 Pennsylvania Avenue in 1942. During World War II, rubber was recycled for use by the Allied Powers in both Europe and Asia. Atlantic Gasoline's billboard was a familiar sight in the area from the 1920s. (Courtesy of the Delaware Public Archives.)

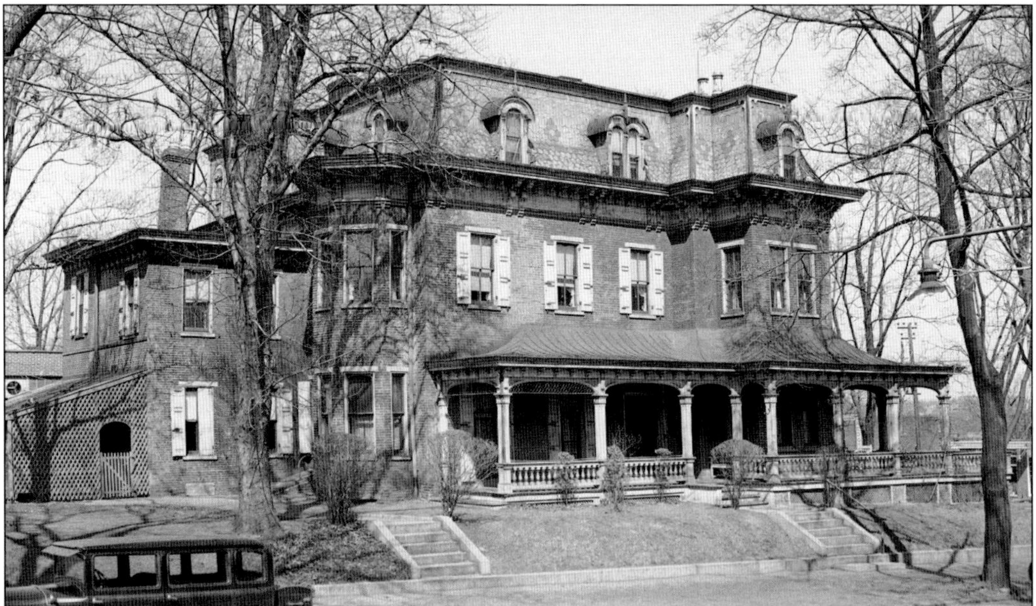

The Gause House was located at Pennsylvania Avenue and Clayton Street in Wilmington. This photograph was taken before the widening of the avenue. Although the house's exact location is uncertain, it was likely on the northwestern corner near what is now Delaware Cadillac. (Courtesy of the Delaware Public Archives.)

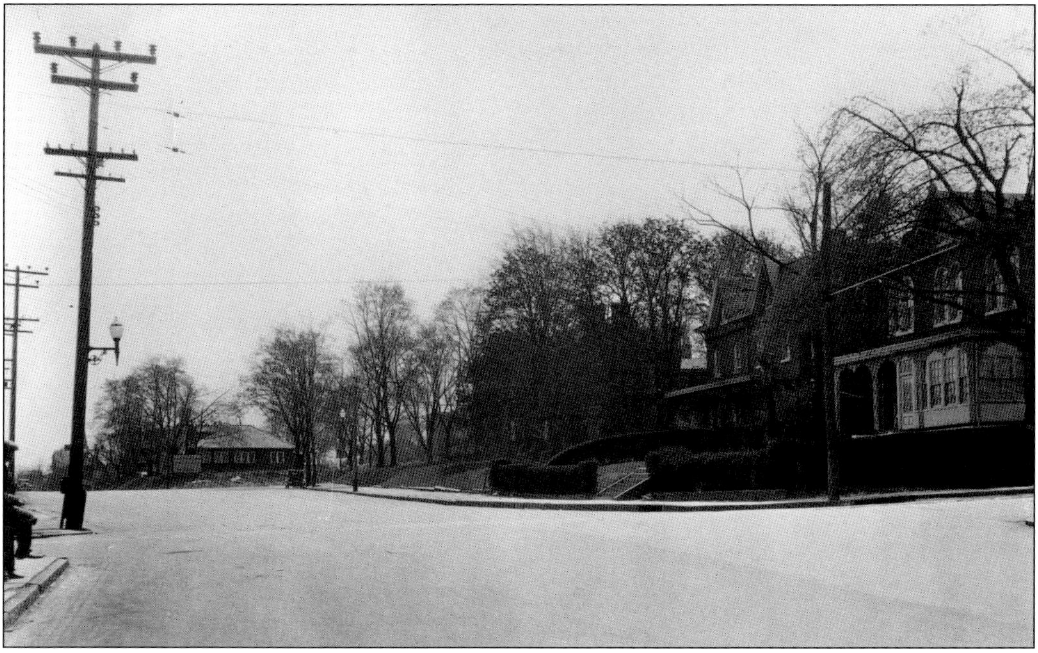

The southeast corner of Pennsylvania Avenue at DuPont Street was just completed when this photograph was taken around 1926. In the distance looking south, one can see the width of the new and old road. Although Pierre S. du Pont bought frontage as it became available, work progressed block by block. (Courtesy of Hagley Museum and Library.)

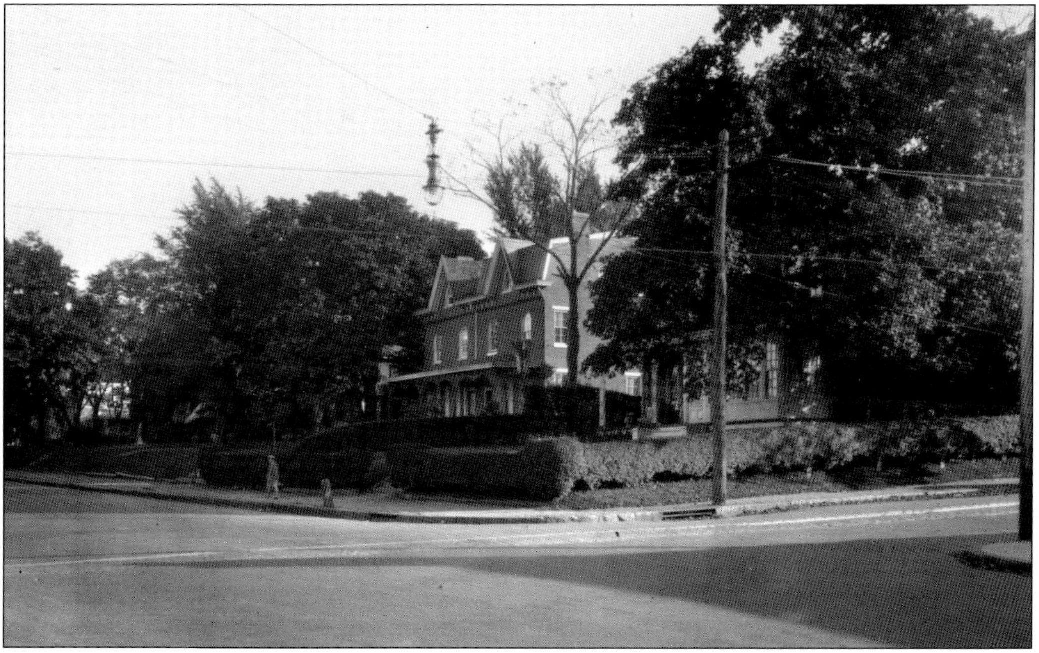

Similar in both location and year to the photograph above, this view of Pennsylvania Avenue at Clayton Street shows the width of Clayton Street at its intersection with the main road. This is likely the site of a present-day office building at 1508 Pennsylvania Avenue. (Courtesy of Hagley Museum and Library.)

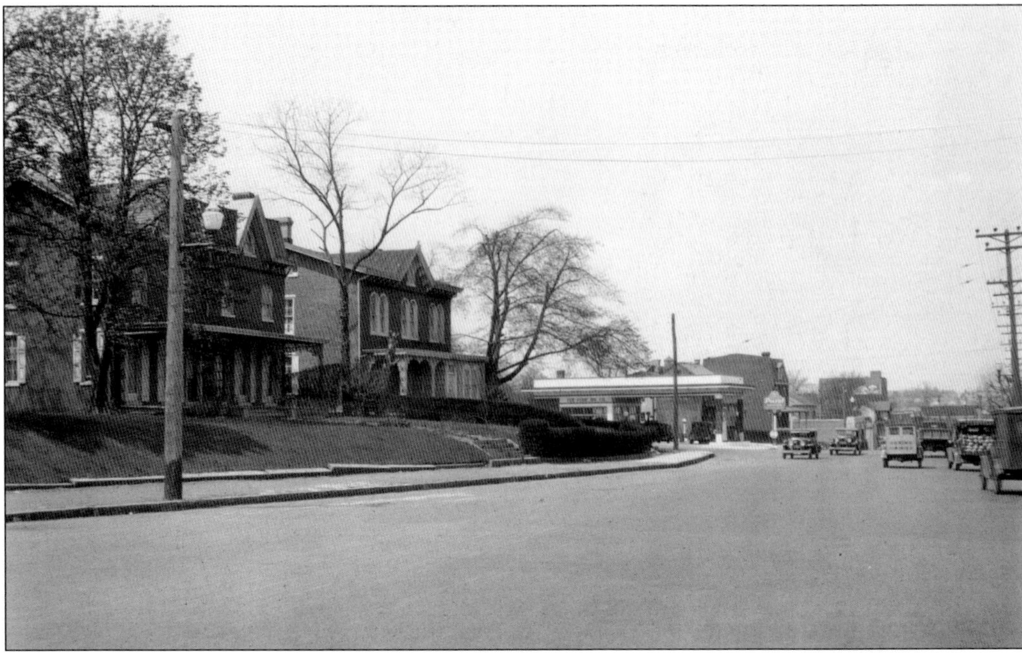

This is another view in the vicinity of Clayton Street, this time focusing to the north and showing the width of Pennsylvania Avenue after construction, around 1927. A Purol service station can be seen north of Clayton Avenue on the corner. Row homes occupy the site of what is now a Cadillac dealership. (Courtesy of Hagley Museum and Library.)

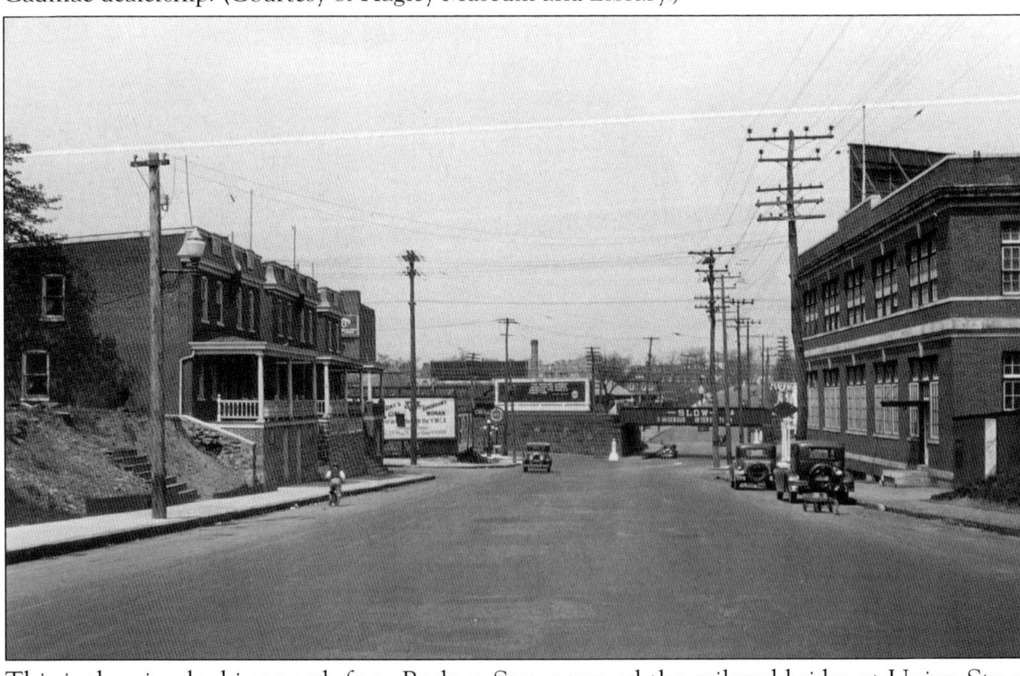

This is the view looking north from Rodney Street toward the railroad bridge at Union Street before Pennsylvania Avenue was widened in 1925. A Gulf service station can be seen at left at the site of the present-day Union Park BMW. The crossing of Lincoln Street and Pennsylvania Avenue can also be seen just before the gas station. (Courtesy of Hagley Museum and Library.)

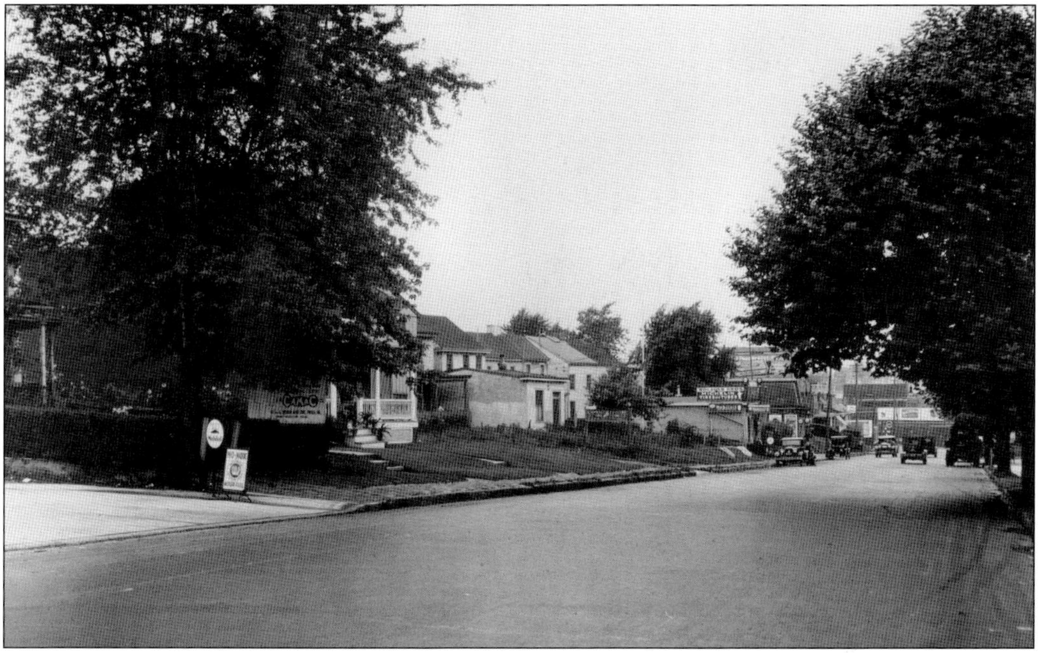

In a similar view to that on the facing page, this c. 1925 photograph shows the homes along Pennsylvania Avenue in the area of Scott and Lincoln Street before the widening of the avenue. Wolhar Brothers service station can be seen at left, and the railroad bridge at Union Street is visible in the background. (Courtesy of Hagley Museum and Library.)

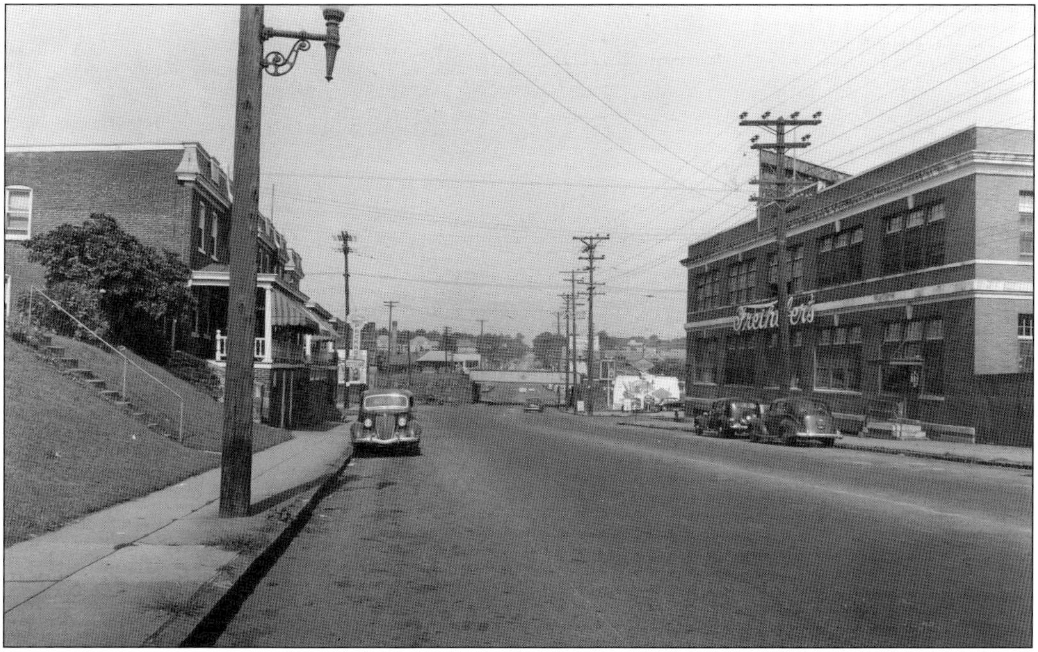

This was the view of Union Street and Pennsylvania Avenue looking north near the railroad overpass in August 1947. Freihoffer's Bakery can be seen at right, as can a Sunoco service station that has since been demolished. Note the two-way traffic on Union Street. (Courtesy of the Delaware Public Archives.)

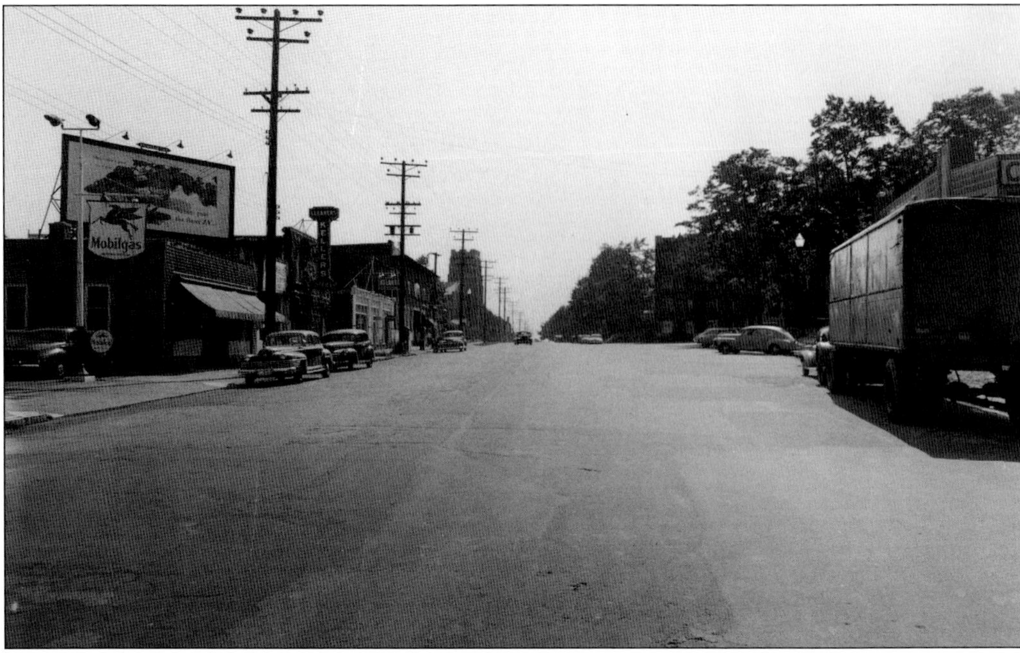

In this view looking south from the railroad overpass near Union Street are one of the Union Park dealerships (likely Pontiac) at right and the Atlantic gas station at left. Pennsylvania Avenue at this vicinity in 1947 was wider to accommodate vehicle parking. (Courtesy of the Delaware Public Archives.)

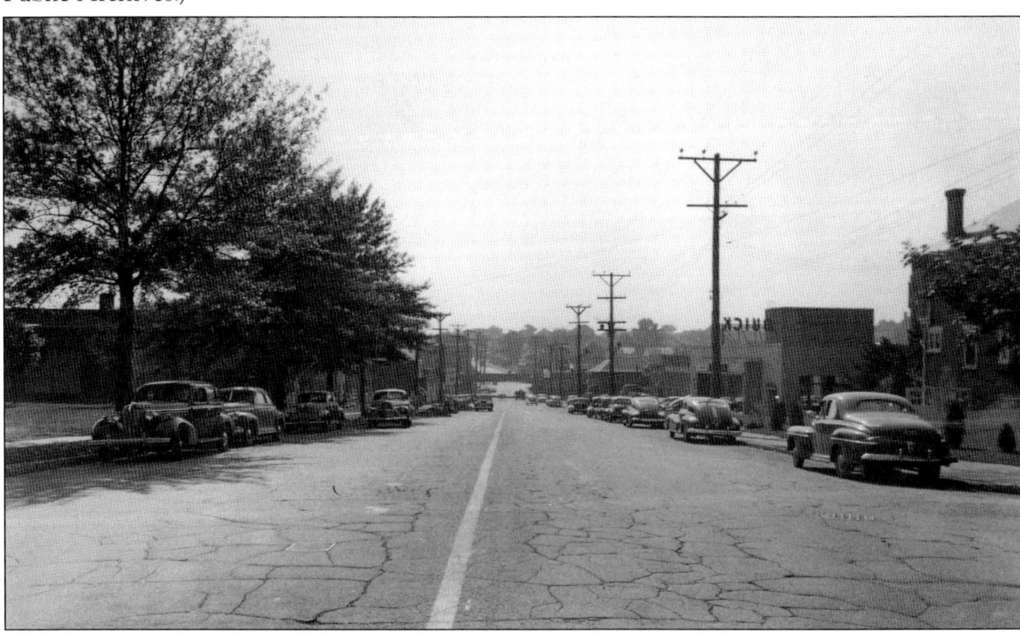

Looking south from about the vicinity of Bancroft Parkway in 1947, one would have seen Liberty Buick at right; farther down the street was Diamond Ice and Coal Company. The car dealership is located approximately at the location of what is today Koons Lexus of Wilmington. Mack's Texaco is barely visible at the corner of Union Street and Pennsylvania Avenue, just before the overpass on the left. (Courtesy of the Delaware Public Archives.)

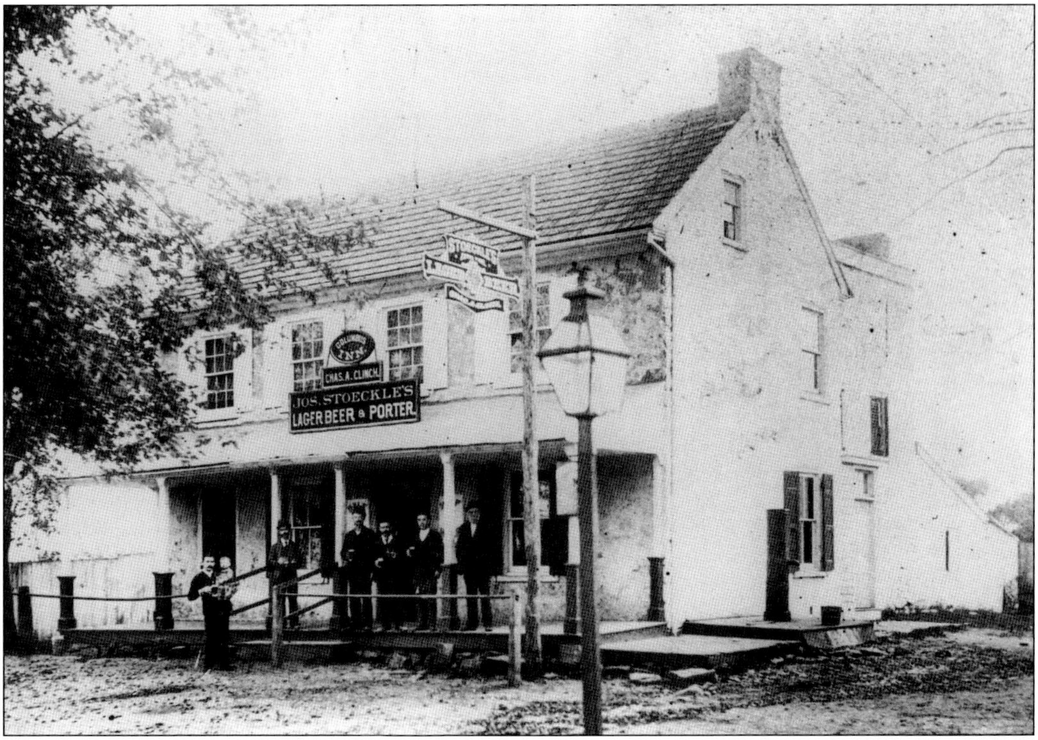

The Columbus Inn, shown above in 1885 and below in 1940, has long been a popular watering hole in the area. It was operated as early 1798 as a bakery by the Schmaltz family, and Elizabeth Schmaltz was a shareholder in the Wilmington & Kennett Turnpike Company. Local legend suggests that she objected to the original placement of the tollgate at the site of the B&O railroad crossing because without side streets, those wishing to stop in for a drink would have to pay a toll. Whether or not this was the reason for the movement of the tollgate to the vicinity of Brecks Lane is unknown. However, the Columbus Inn has operated as a tavern since 1812 and assumed its current name by 1849. It was operated by Wallace Sezna from 1953 to 2007 and is now run by a member of the Capano family. (Both, courtesy of Hagley Museum and Library).

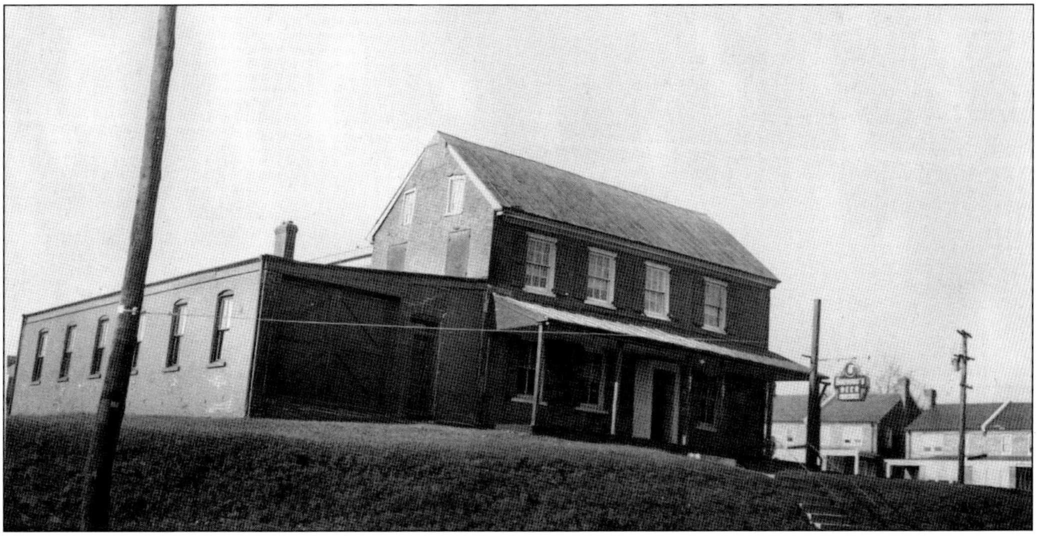

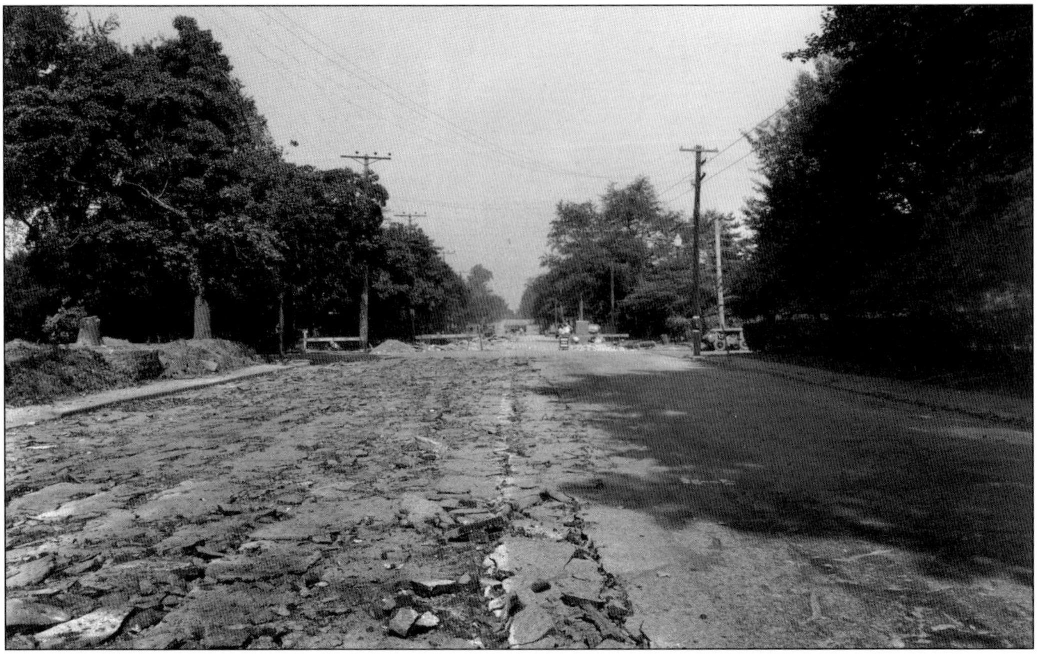

Shown here is Pennsylvania Avenue looking north from Wilmington at the intersection with Greenhill Avenue in 1947. By this time, the road was undergoing another resurfacing. Today, the Devon apartment building is at the right, and "Gibraltar," the home of H. Rodney and Isabella Sharp, is just across Greenhill Avenue. (Courtesy of the Delaware Public Archives.)

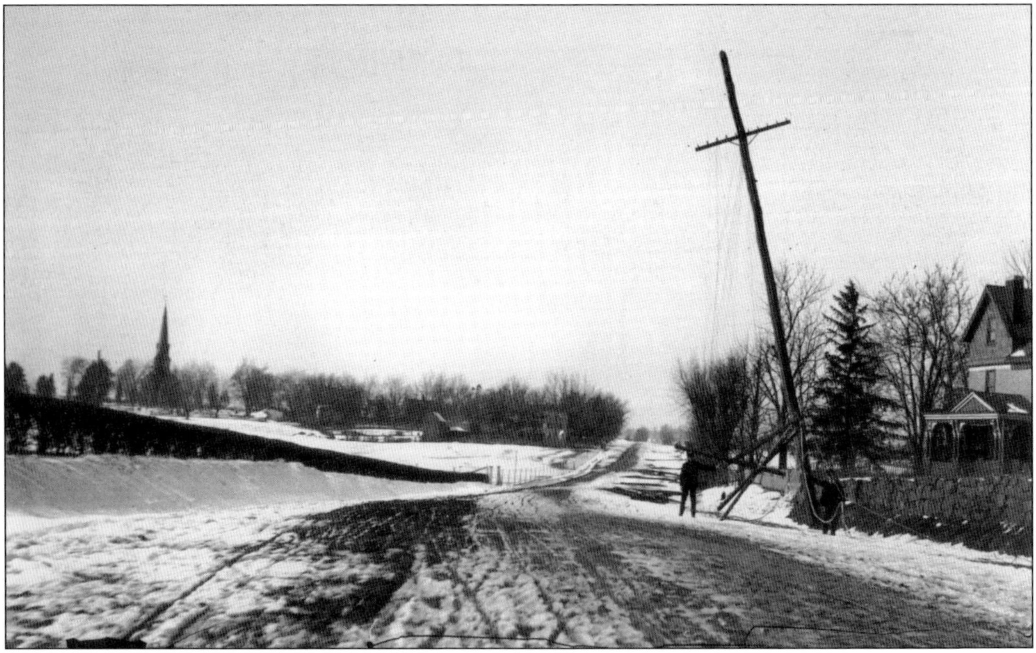

Taken between 1900 and 1919 just a bit farther up from the image above, near the intersection of Kennett Pike with Tower Road, this photograph shows the width of the road before P.S. du Pont undertook the project. The steeple of Greenhill Presbyterian Church can be seen in the distance. (Courtesy of Hagley Museum and Library.)

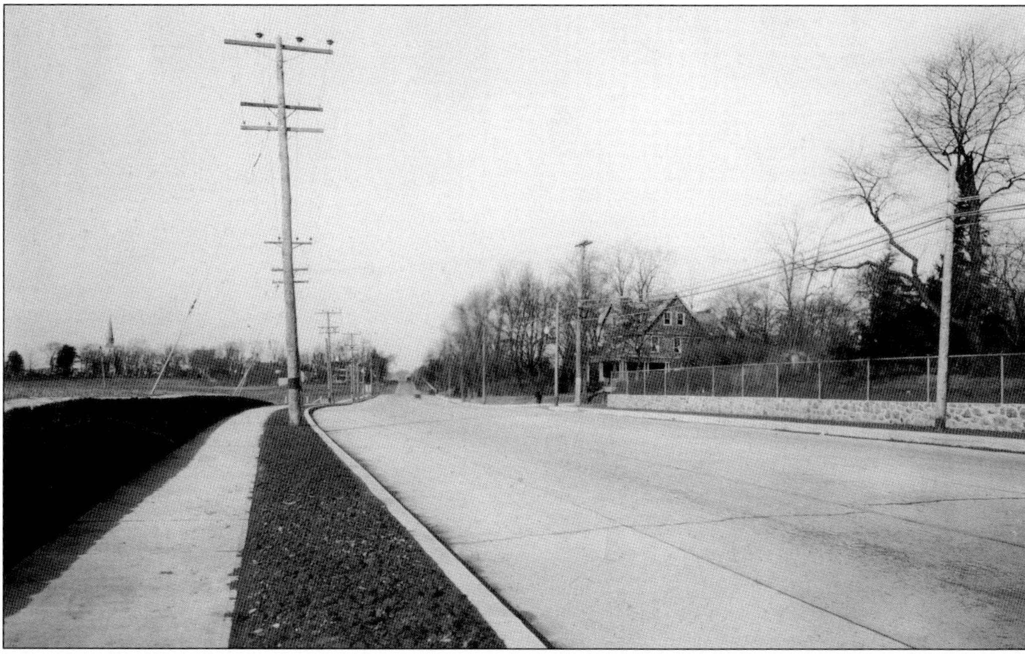

Looking closely, one can see the original width of Kennett Pike in the distance with its soon-to-be-paved sides and, in the foreground, the width of the new pike. Taken right after this section was completed around 1921, in the vicinity of Mount Salem Lane, the stone wall at right encompasses Goodstay, a du Pont family residence. (Courtesy of the Delaware Public Archives.)

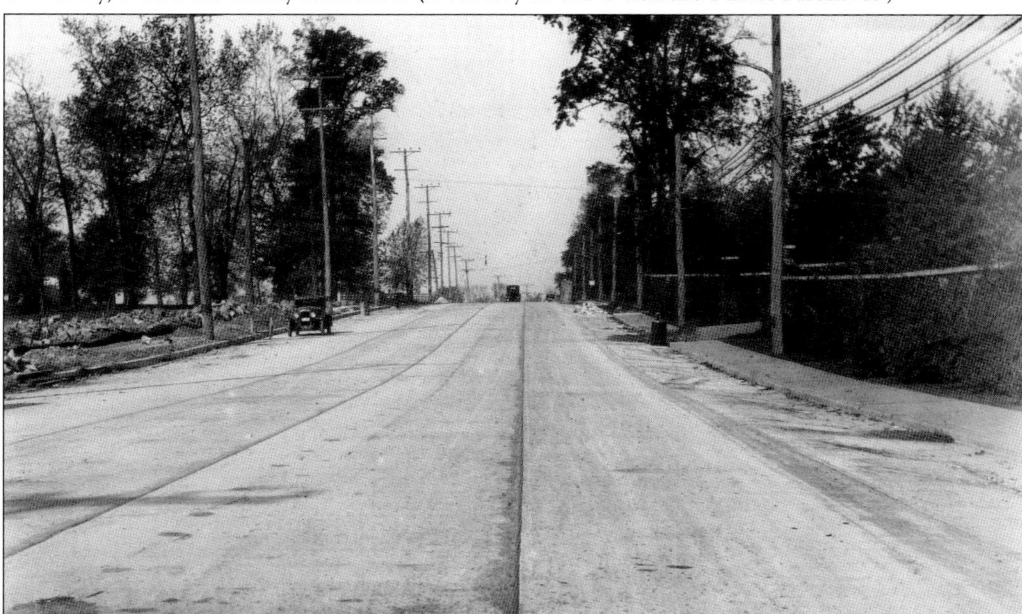

This photograph captures Kennett Pike just north of Greenhill Avenue not long after the widening of the pike, around 1920. At left, one can see the wall being built between Kennett Pike and Goodstay, which is now the Wilmington Campus of the University of Delaware. At right is the home of P.S. du Pont's sister Isabella and her husband, H. Rodney Sharp. (Courtesy of the Delaware Public Archives.)

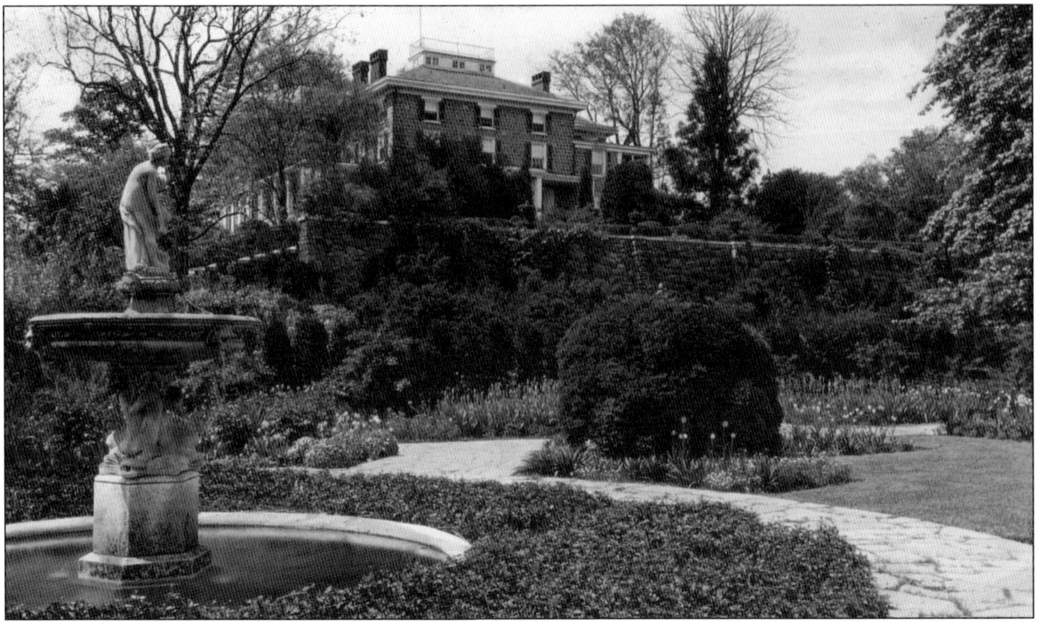

Gibraltar was the home of H. Rodney Sharp and Isabella Sharp, the sister of Pierre S. du Pont. Shown here in 1943, the home features a Marian Coffin–designed garden. The original house was built in 1844 and expanded several times under the ownership of the Sharps. Bordering both Kennett Pike and Greenhill Avenue, the estate has long stood as a Wilmington landmark. (Courtesy of Hagley Museum and Library.)

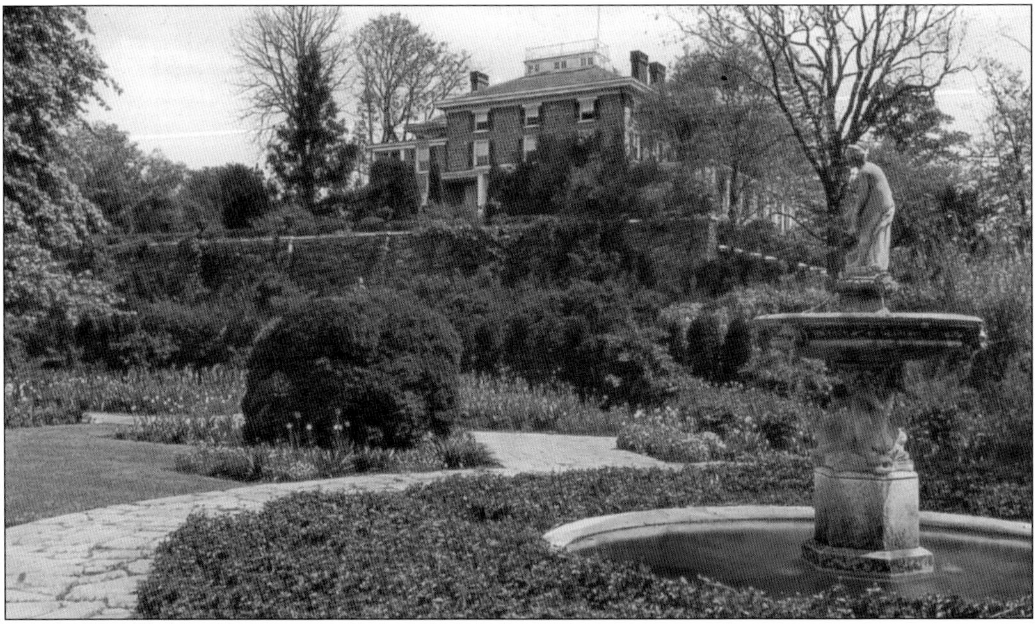

The Lower Garden at Gibraltar was one of the many terraced levels of the garden at the estate. This view from approximately 1925 shows in greater detail the lavishness of the Coffin-designed garden. Marian Coffin is well known as one of the first and perhaps most-accomplished female landscape architects in the United States. The gardens of this 6.1-acre property is open for visitation. (Courtesy of Hagley Museum and Library.)

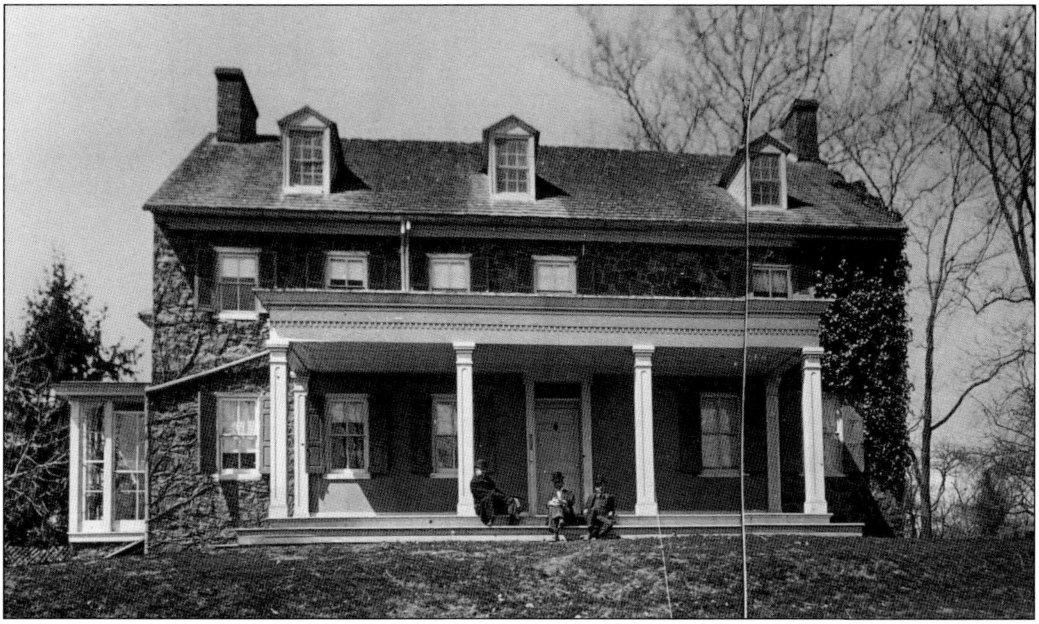

Located at 2600 Pennsylvania Avenue is Goodstay, originally known as Green Hill. Dating to 1740 and shown in April 1896, the home was later enlarged by Ellen Meeds Wheelwright, daughter of T. Coleman du Pont. Ellen Wheelwright and her husband, Robert, are also credited with the creation of the gardens. The property now forms the Wilmington Campus of the University of Delaware. (Courtesy of Hagley Museum and Library.)

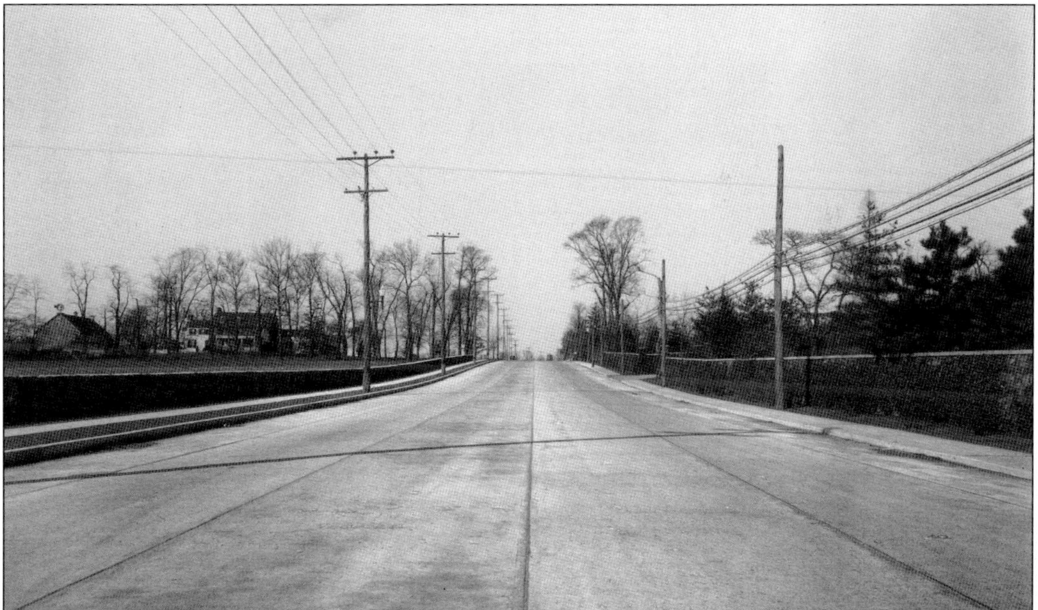

This view is looking north on Kennett Pike, just past Greenhill Avenue, after the widening of the pike by Pierre S. du Pont in the 1920s. This view, taken in approximately 1930, shows the width of the road being nearly the same as it is today, with Goodstay visible at left, and Gibraltar, at right. (Courtesy of the Delaware Public Archives.)

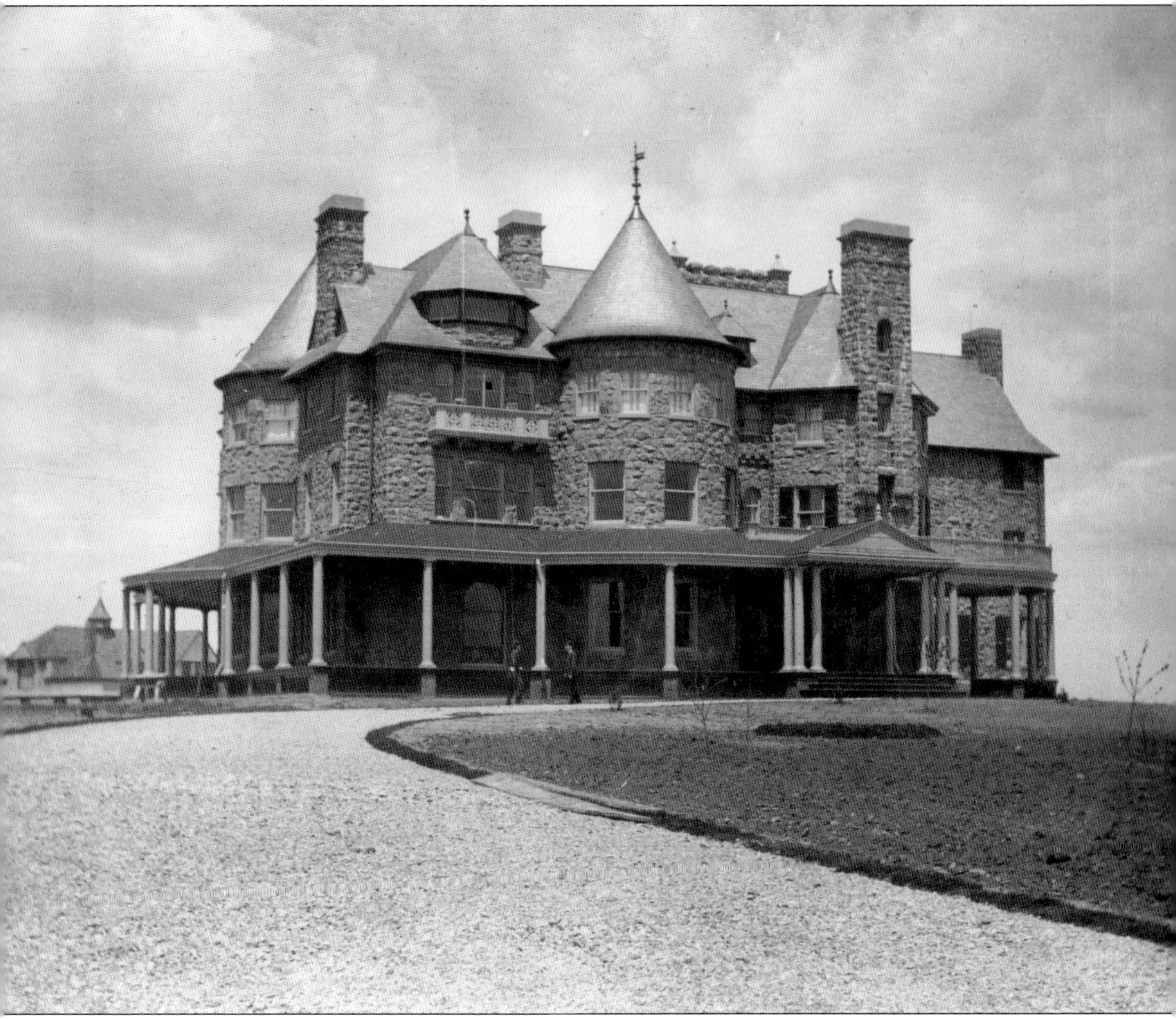

St. Amour, the home of Mary Belin du Pont (1839–1913), is pictured here around the turn of the 20th century. St. Amour was the first of seven du Pont family homes constructed on what was known as "the Neighborhood." It was built in 1892 following the construction of A.I. du Pont School in 1890. The home featured a formal garden, tennis court, and service/garage complex. After the death of Mary du Pont in 1913, the home was occupied by her son Lammot du Pont (1880–1952). In addition to the luxuries of tennis courts and lavish gardens, the Neighborhood also featured a community swimming pool on the part of the complex that was home to the A. Felix du Pont house known as Elton. Built in the days before modern pool equipment and chemicals were readily available, the pool was drained and refilled every Thursday night. Copper sulfate was added to keep the water free of algae. (Courtesy of Hagley Museum and Library.)

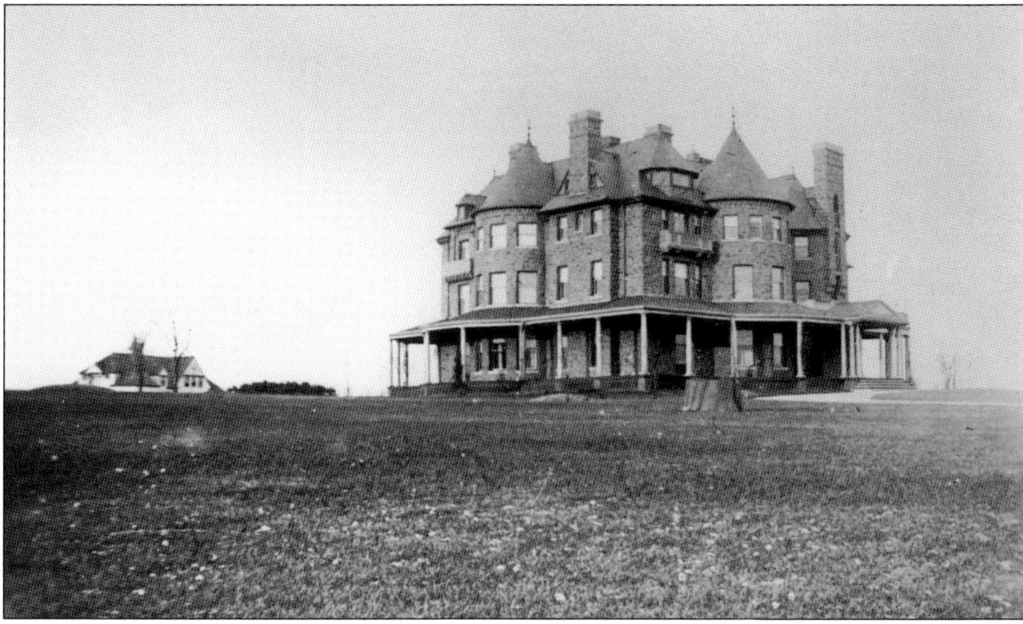

Shown is St. Amour shortly after construction in 1892. Pierre S. du Pont, financier of the Kennett Pike project, moved to this home as a teenager following the death of his father in 1884. His mother, Mary Belin du Pont, returned to Wilmington to this house to raise her 10 children in the company of other family members. (Courtesy of Hagley Museum and Library.)

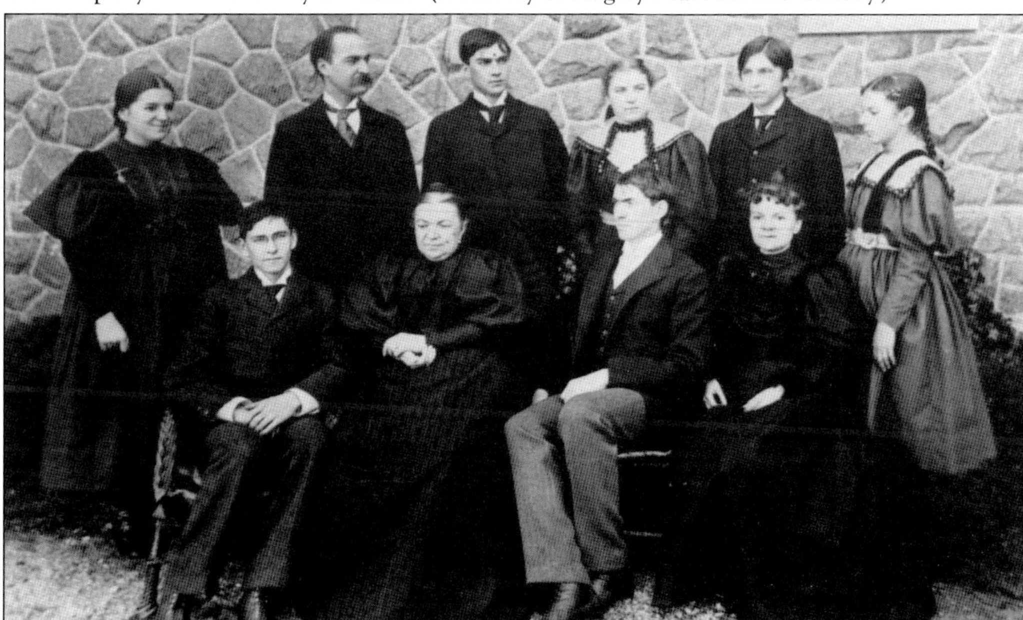

This family portrait shows the original occupants of St. Amour gathered in 1894 around their mother, Mary. From left to right are (first row) Irénée, Mary Belin du Pont, Henry Belin, and Louisa (later Copeland); (second row) Isabella (later Sharp), Pierre S. (of Longwood), William K., Mary (later Laird), Lammot, Margaretta (later Carpenter). (Courtesy of Hagley Museum and Library.)

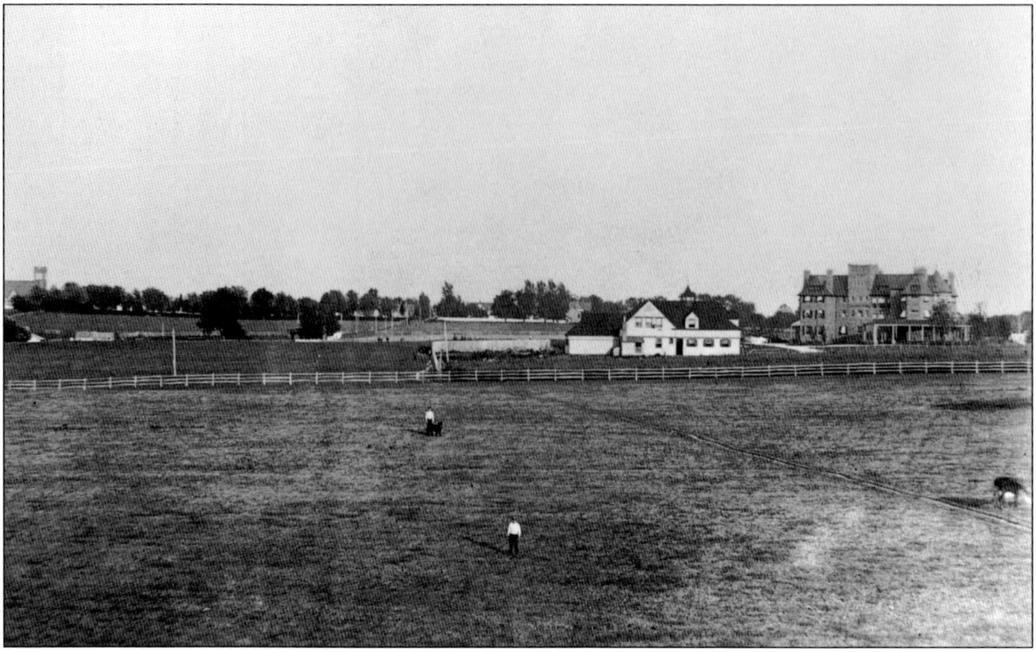

This image of St. Amour and the barn was taken by Pierre Gentieu, a DuPont Company employee, who photographed the Brandywine region around the powder mills. Taken in 1893 from the scaffolding of Alexis I. du Pont School, it shows the enormous size of the home as well as its open surroundings. (Courtesy of Hagley Museum and Library.)

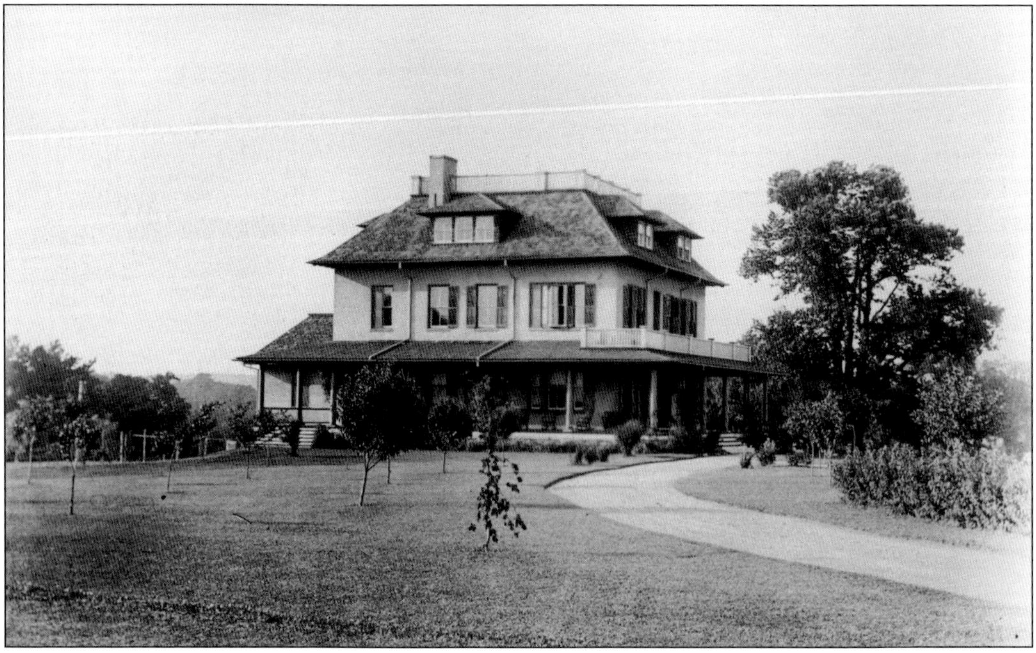

This home, known as Square House and pictured around 1905, was built for Mary Belin du Pont's son Irénée and his wife, Irene. The home was sold in 1923 to Walter Carpenter Jr., the first non-du Pont at the head of the family firm. Carpenter enlarged the home and renamed it St. Giles. (Courtesy of Hagley Museum and Library.)

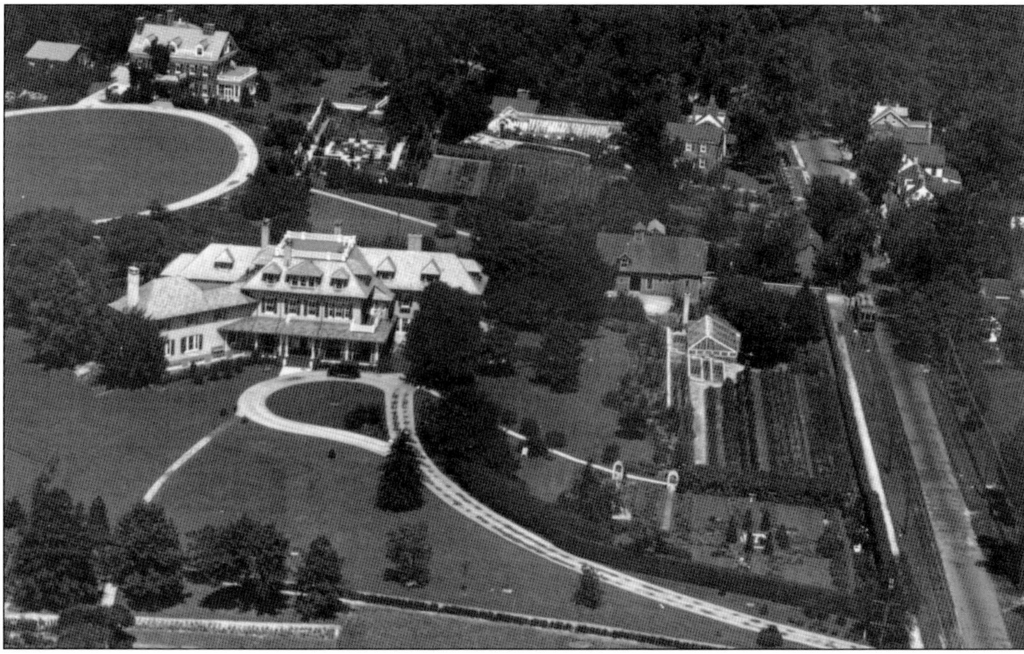

This 1930 aerial photograph shows the enlarged St. Giles as it appeared when the Carpenter family finished with the enlargement of the main house. Rising Sun Lane can be seen at right, and the Copeland House, with its circular driveway, can be seen at left. (Courtesy of Hagley Museum and Library.)

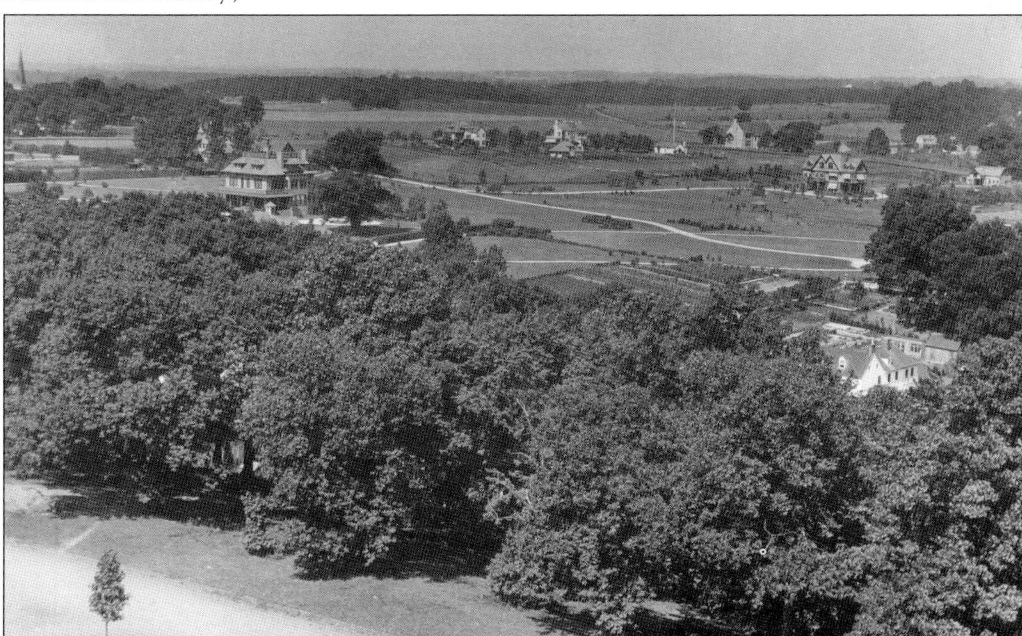

This photograph, taken from Rockford Tower around 1920, shows a few of the houses that made up the Neighborhood. From left to right are St. Giles, Elton, the Ernest du Pont home, Alexis I. du Pont School, and Windmar. Kennett Pike lies just in front of the school. (Courtesy of Hagley Museum and Library.)

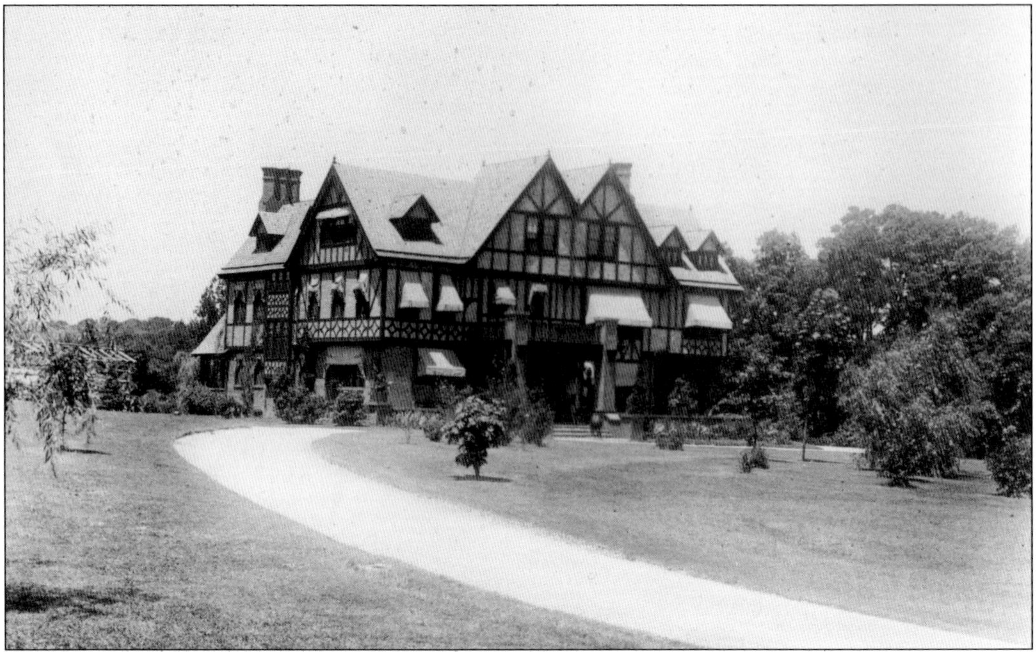

The home of William Winder and Mary Aletta Belin du Pont Laird, known as Windmar, is the only surviving dwelling in the Neighborhood. After the death of Mrs. Laird (a sister of Pierre S. du Pont) in 1938, her daughter Wilhelmina Laird Craven inherited the property. It is still owned by the Craven family. (Courtesy of Hagley Museum and Library.)

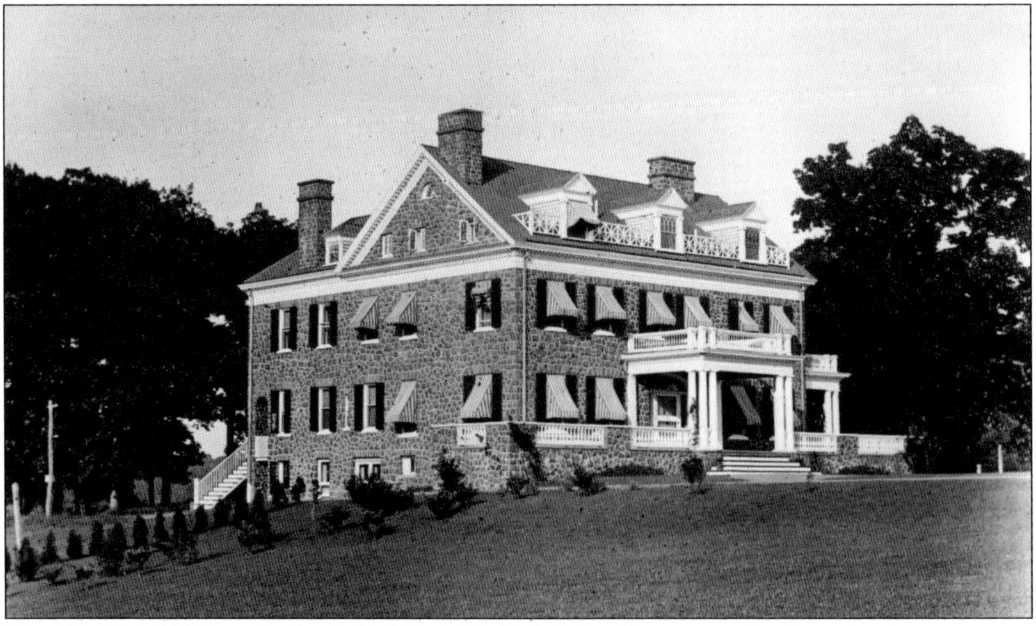

The Copeland House was built for Pierre S. du Pont's oldest sister, Louisa, after her marriage to Charles Copeland in 1904. Louisa Copeland died in 1926, leaving her husband on his own. After her untimely death, her cousin Aileen du Pont became the housekeeper and hostess. The home was demolished in 1965 and replaced with a more modern dwelling. (Courtesy of Hagley Museum and Library.)

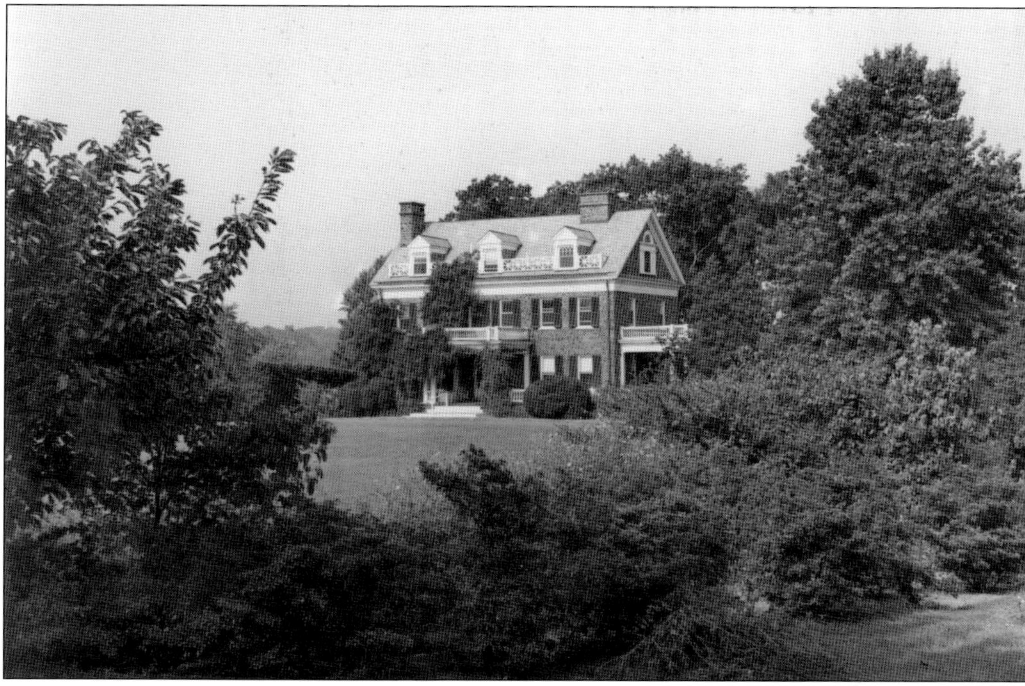

The Copeland House also featured a formal garden, as did all of the others in its vicinity. Every family in the Neighborhood owned a vehicle called the Auto Red Bug. The one owned by Felix du Pont caught fire on the lawn of the Copeland House. (Courtesy of Hagley Museum and Library.)

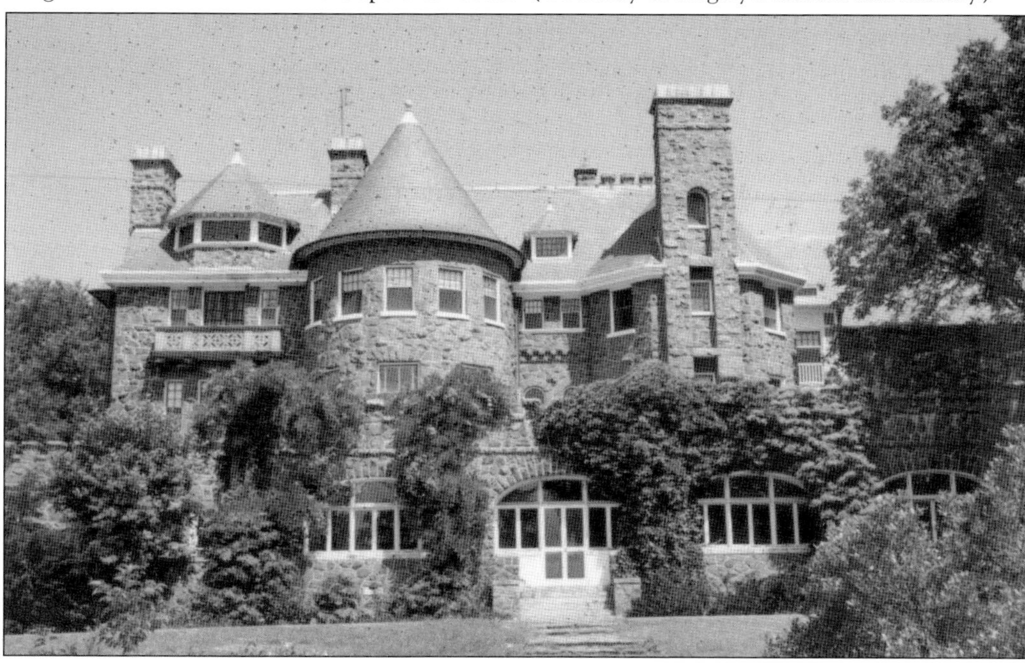

St. Amour shown, here in 1970, had been the centerpiece of the tight-knit community of various du Pont family members. After the passing of former DuPont Company president Lammot du Pont in 1952, his widow, Margaret, lived in the stone mansion at the corner of Rising Sun Lane and Kennett Pike. (Courtesy of Hagley Museum and Library.)

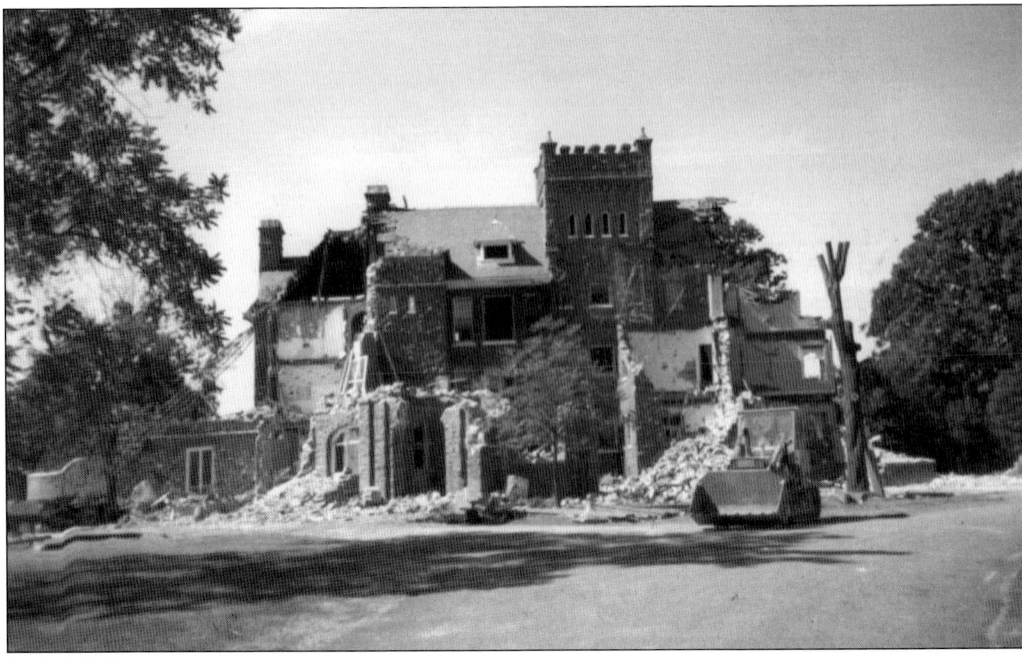

St. Amour met the same fate as St. Giles and was demolished in 1969 after the death of Lammot du Pont's fourth wife. The property was given to Tower Hill School to make way for athletic facilities. Today, the garage and walls of the Marian Coffin garden are all that remain of this centerpiece of the du Pont family. (Courtesy of Irénée du Pont Jr.)

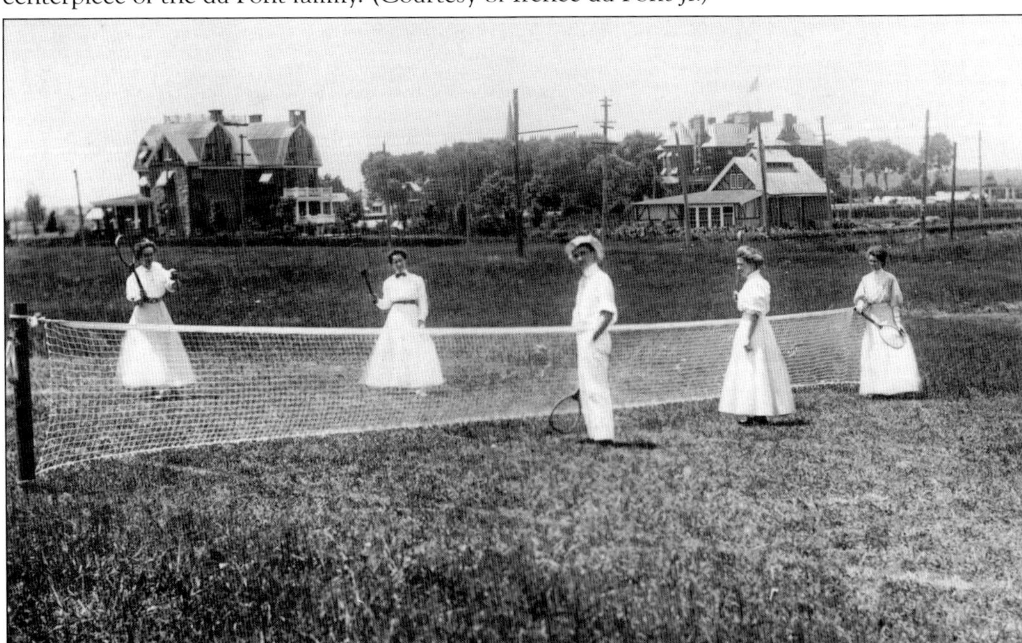

In this c. 1910 photograph, one can see St. Amour at right, and at left, the Victor du Pont House. Known as "Fat Vic," Victor du Pont was in charge of the company's sales representatives. The players on the field are unidentified. Kennett Pike ran between the two homes. (Courtesy of Hagley Museum and Library.)

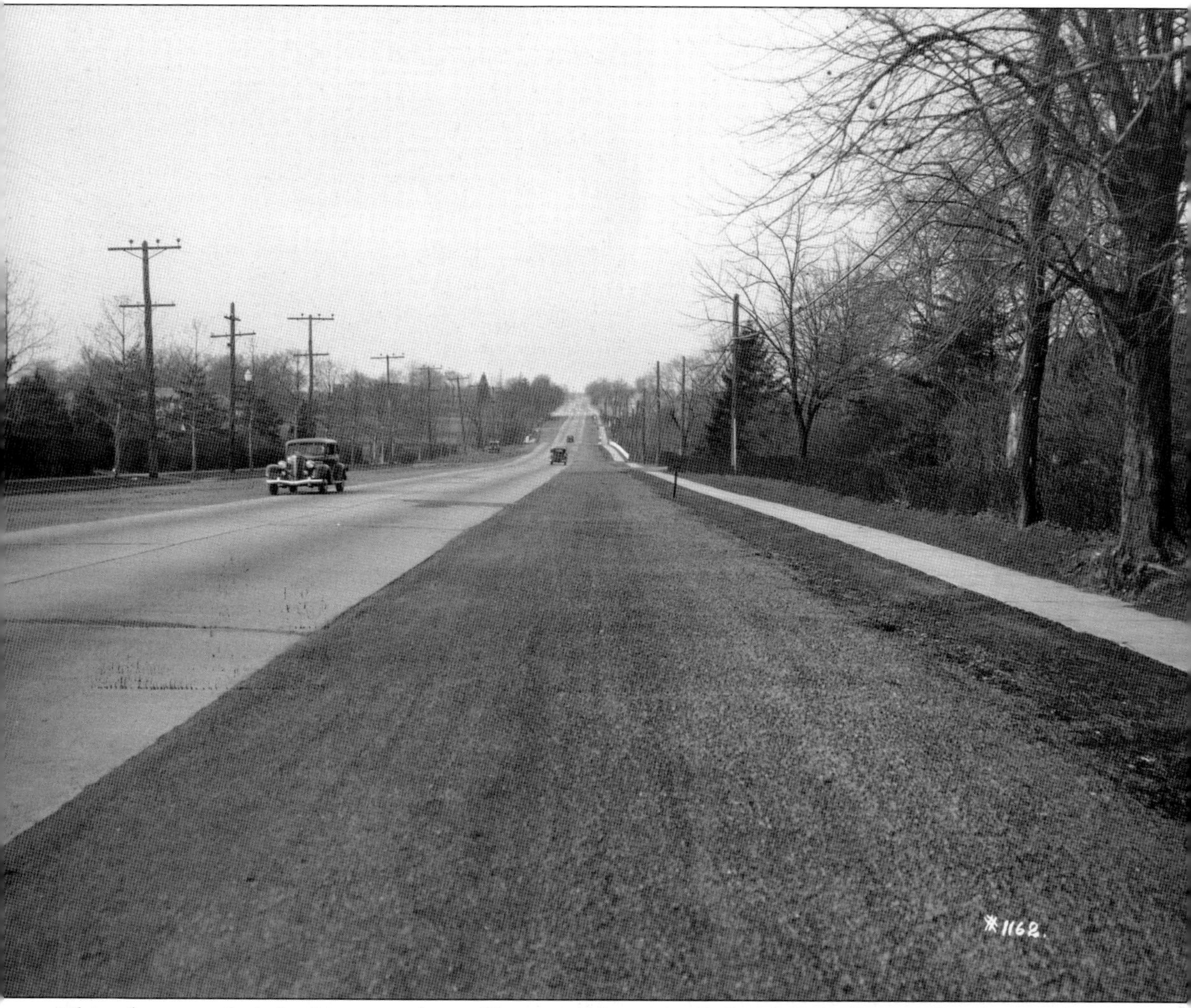

This photograph is looking north on Kennett Pike toward the A.I. du Pont School near the driveway to Windmar in 1934. Viewers can see the paved road as well as the wide right-of-way provided by Pierre S. du Pont. William du Pont's planned affluent suburb of Westover Hills is at left. (Courtesy of the Delaware Public Archives.)

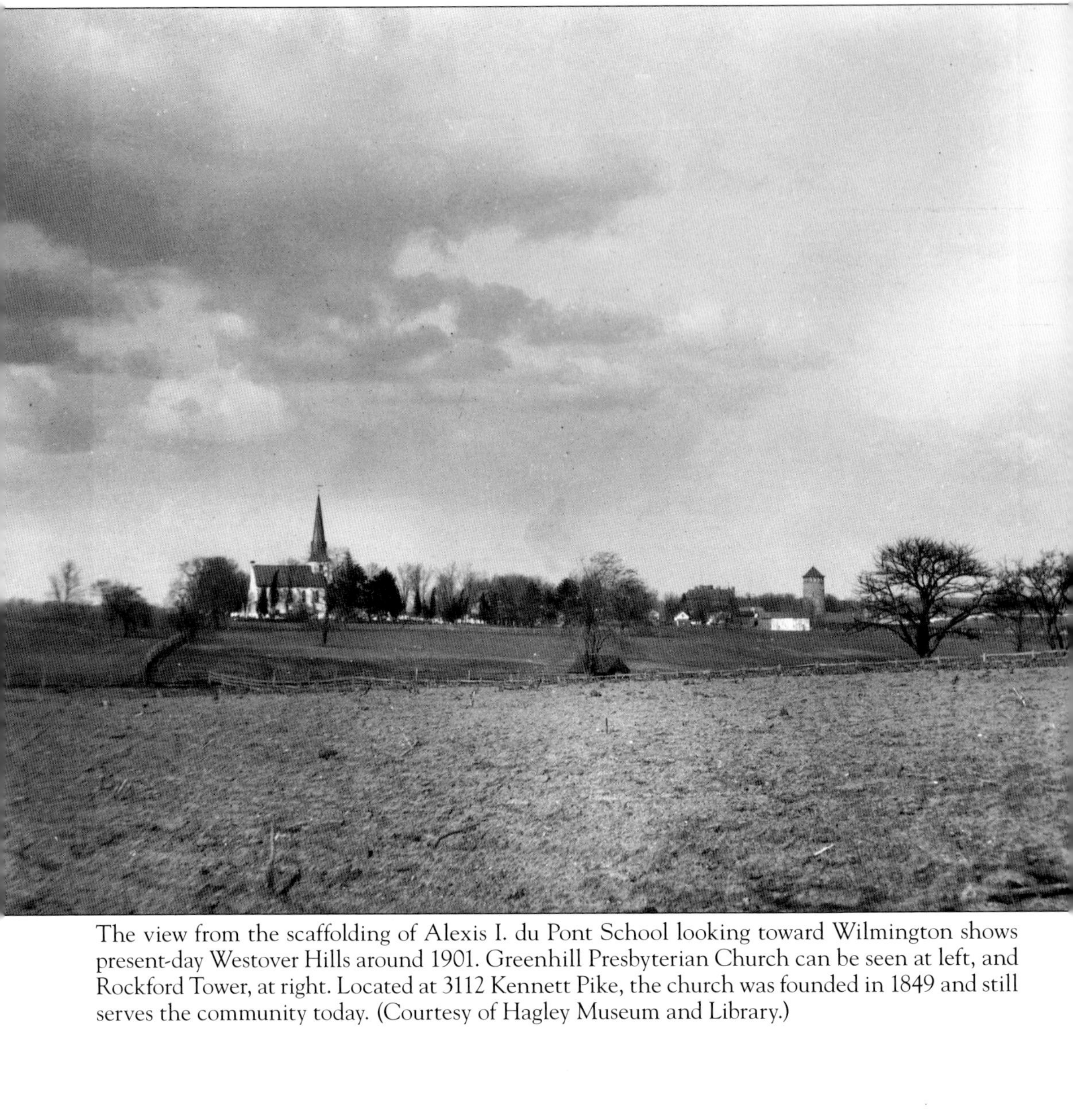

The view from the scaffolding of Alexis I. du Pont School looking toward Wilmington shows present-day Westover Hills around 1901. Greenhill Presbyterian Church can be seen at left, and Rockford Tower, at right. Located at 3112 Kennett Pike, the church was founded in 1849 and still serves the community today. (Courtesy of Hagley Museum and Library.)

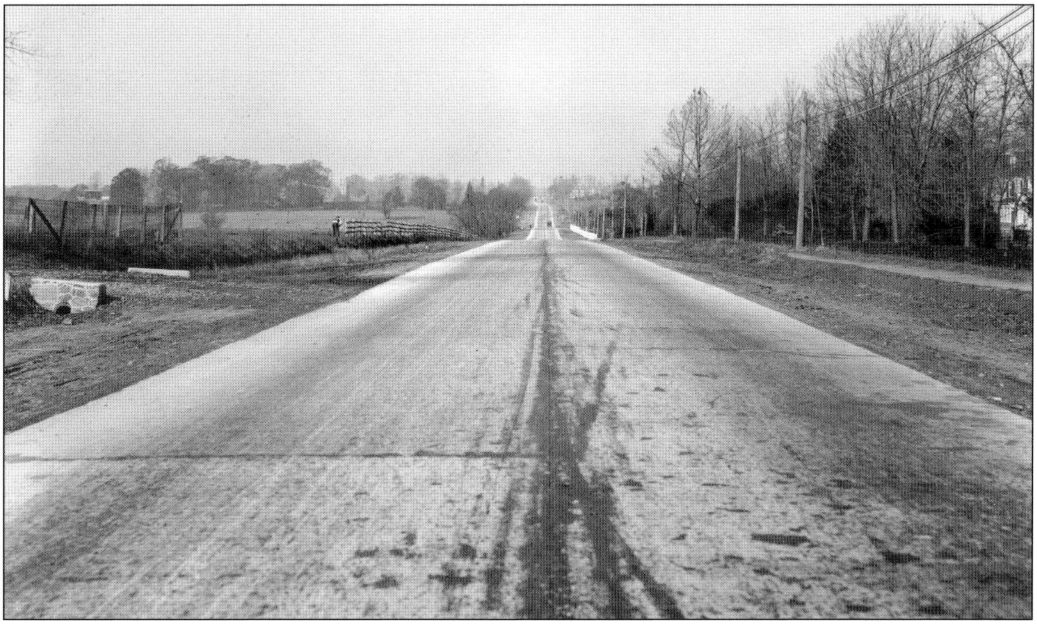

This is another view of Kennett Pike looking north around 1920. The white guardrail in the distance at right is the railroad overpass just before Brecks Lane (a modification made by du Pont from a grade crossing). The dwelling in the distance is the William Raskob House before renovations changed its appearance to what travelers see today. (Courtesy of Hagley Museum and Library.)

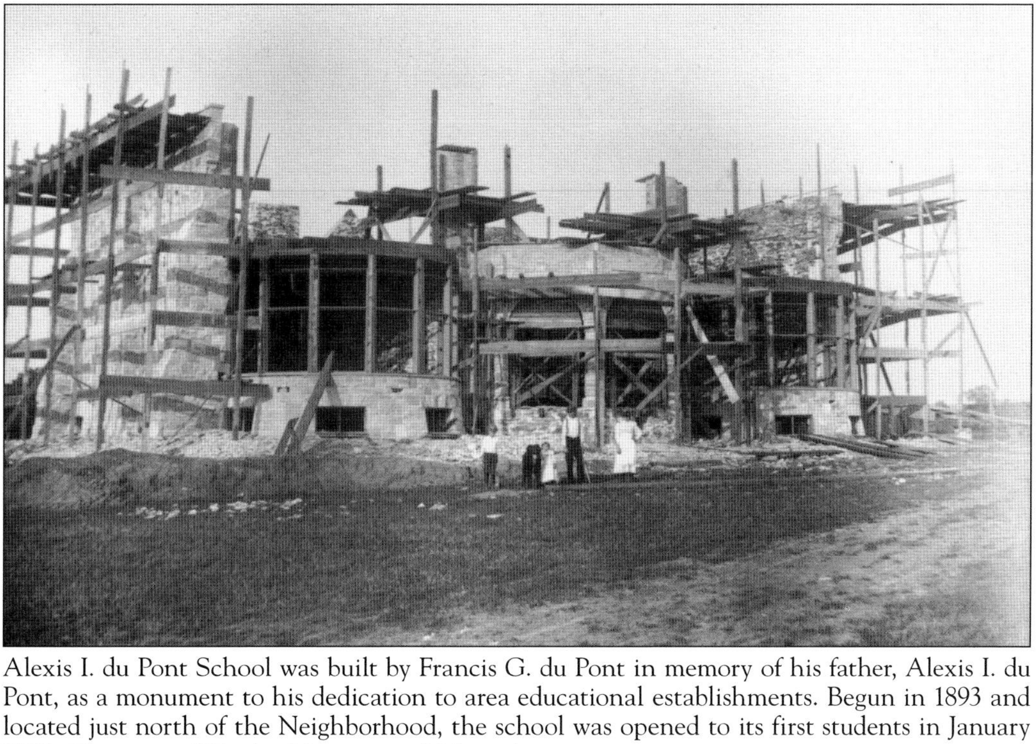

Alexis I. du Pont School was built by Francis G. du Pont in memory of his father, Alexis I. du Pont, as a monument to his dedication to area educational establishments. Begun in 1893 and located just north of the Neighborhood, the school was opened to its first students in January 1894. (Courtesy of Hagley Museum and Library).

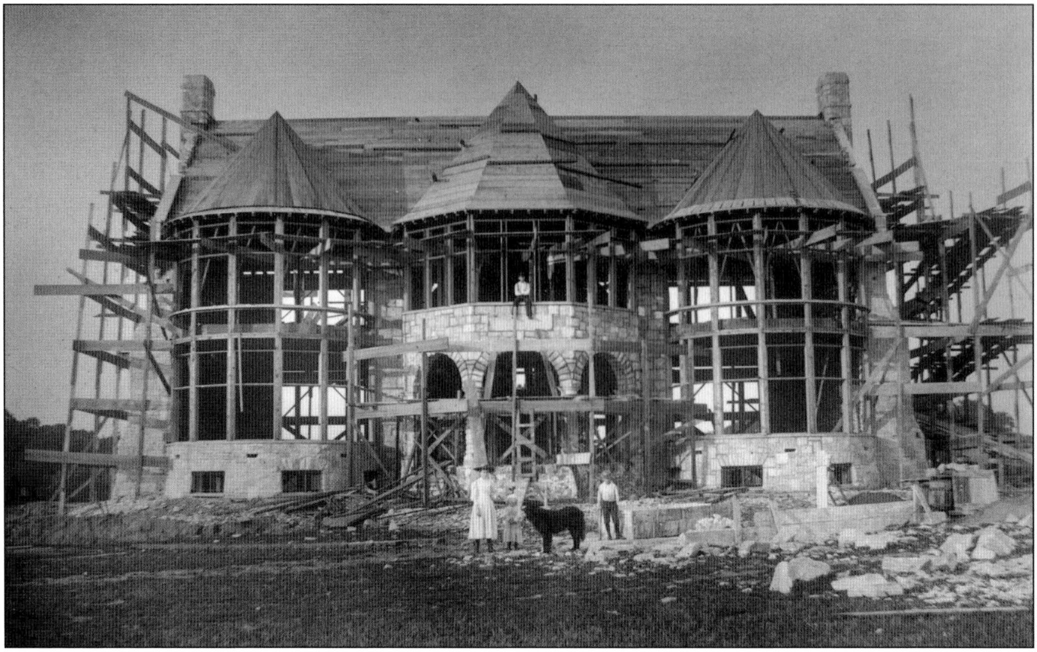

Another view of A.I. du Pont School under construction in 1893 shows masonry work beginning on the facade of the building. Fondly referred to as "A.I." by residents, the school replaced two smaller schools nearby that served many children of workers in the DuPont powder mills. The individuals in the foreground are unidentified. (Courtesy of Hagley Museum and Library.)

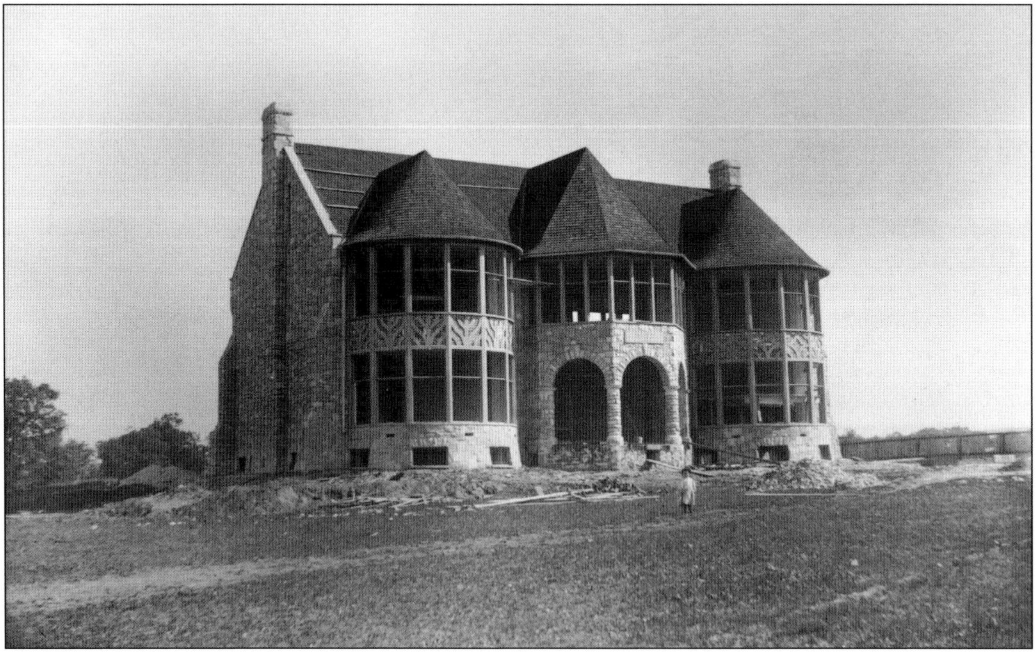

Shown is the completed A.I. du Pont School in late 1893, likely photographed from Kennett Pike. The building has undergone various expansions over the past century; however, it retains this unique facade today. Classes at all grade levels were held here until 1952, when a new elementary school was opened in Greenville. (Courtesy of Hagley Museum and Library.)

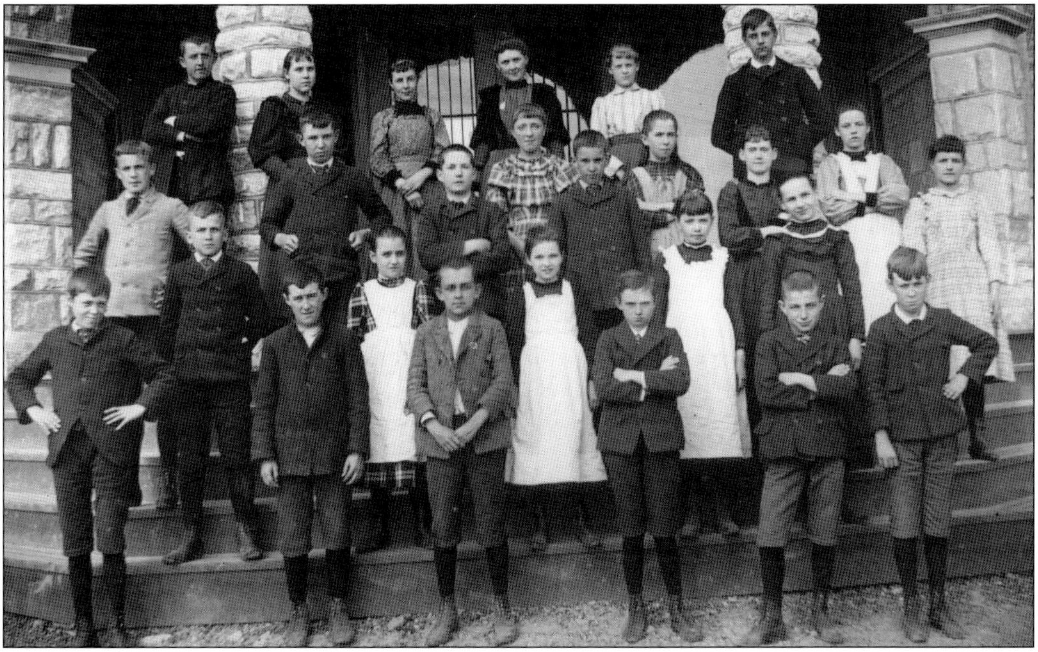

This photograph taken by Pierre Gentieu, Brandywine photographer and employee of the powder mills, shows the class of Miss Wither in 1894 on the school's front steps. While none are identified, these were among students to attend the new school built by Francis G. du Pont. (Courtesy of Hagley Museum and Library.)

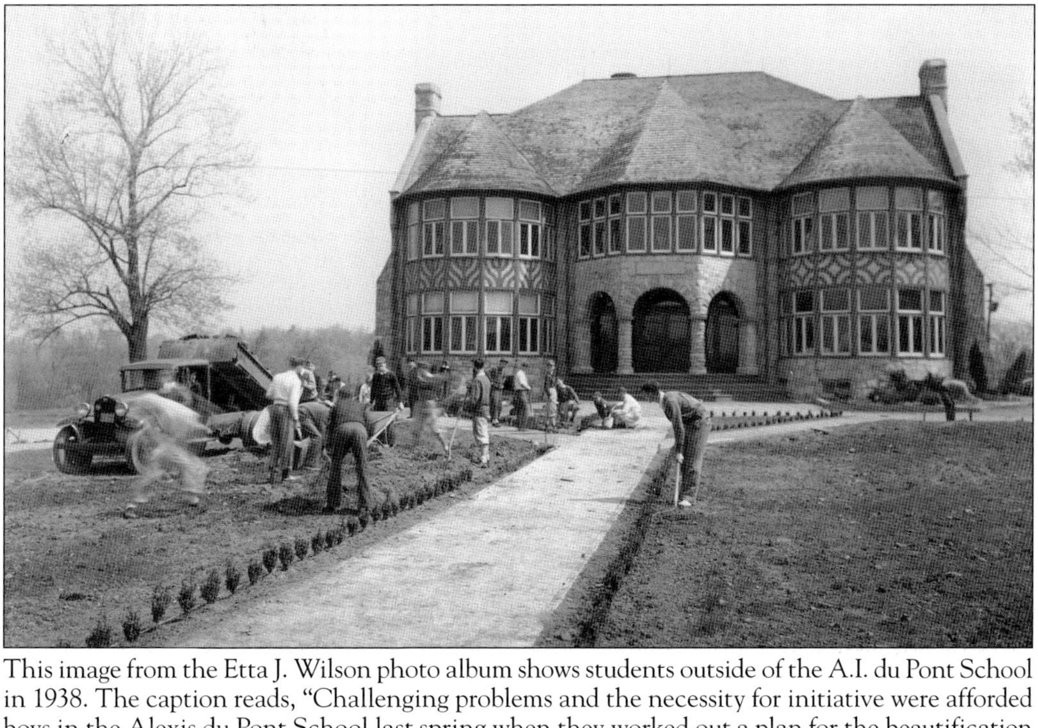

This image from the Etta J. Wilson photo album shows students outside of the A.I. du Pont School in 1938. The caption reads, "Challenging problems and the necessity for initiative were afforded boys in the Alexis du Pont School last spring when they worked out a plan for the beautification of the school grounds on the Kennett Pike." (Courtesy of the Delaware Public Archives.)

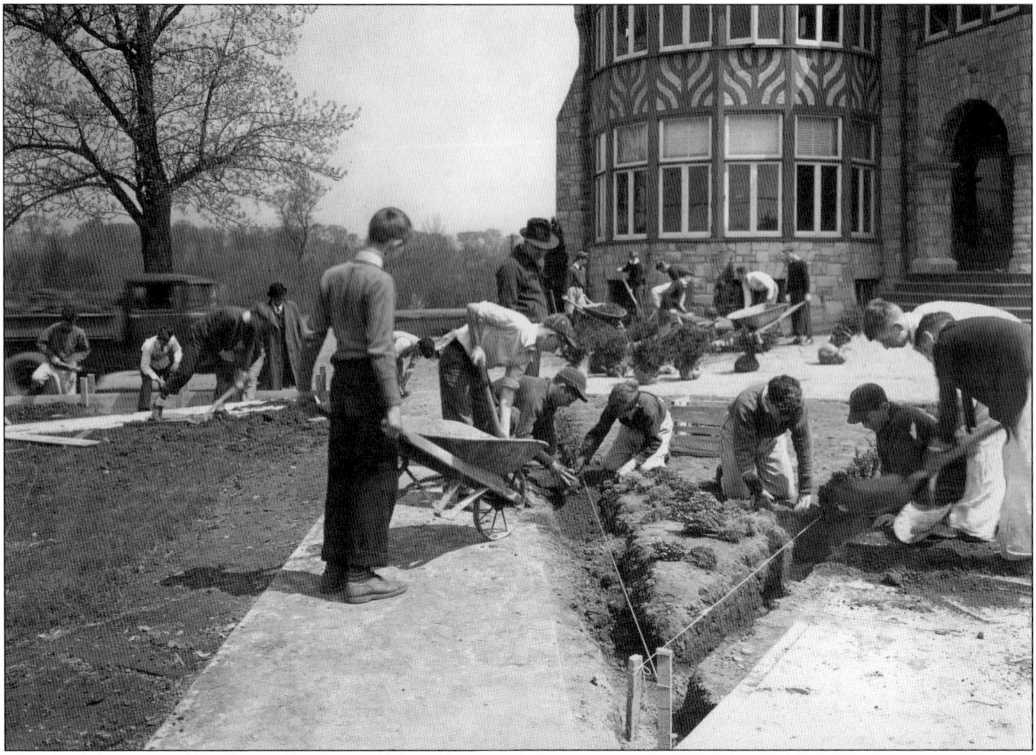

Here, more work is being done by the students of the A.I. du Pont School on beautification of the grounds in 1938. After 72 years, a new Alexis I. du Pont High School was constructed farther north along Kennett Pike in Greenville in 1966. Currently, this building houses Alexis I. du Pont Middle School, part of the Red Clay Consolidated School District. (Courtesy of the Delaware Public Archives.)

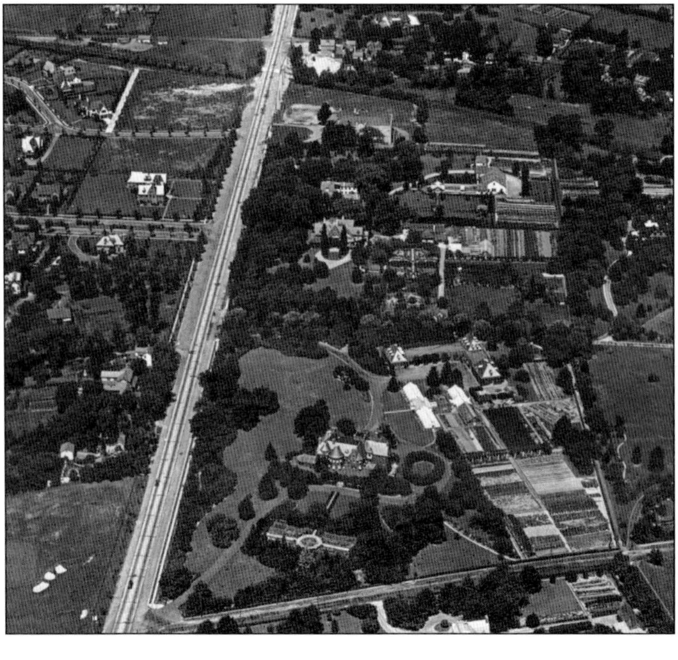

This aerial view of Kennett Pike (center of image) shows the du Pont Family's "Neighborhood" at right, with St. Amour visible at the intersection with Rising Sun Lane (near bottom of image). The first road to the left leads to the upscale neighborhood of Westover Hills. (Courtesy of Hagley Museum and Library.)

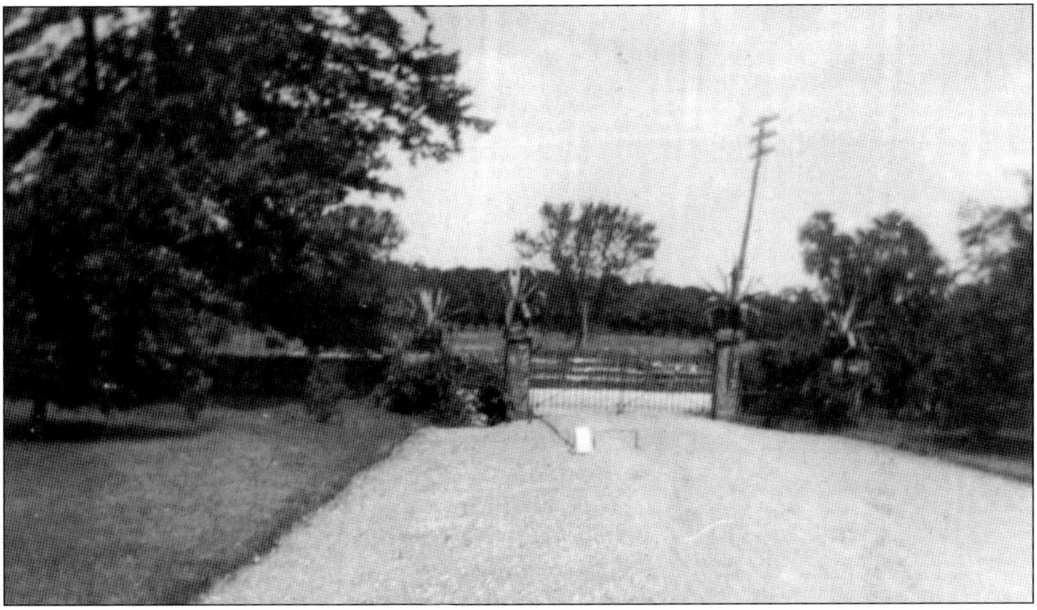

Crossing Route 100 on Kennett Pike led travelers to the gates of this du Pont family estate, Pelleport, pictured here around the turn of the century. The home was built in 1881 by Henry Du Pont for his son William on the occasion of his marriage. Following William's divorce and Henry's death in 1889, the property was vacated. After an explosion in the powder yards in 1890, William's mother, Louisa, and sister Evelina moved into the house after their own was nearly destroyed. Louisa died in 1900, and Evelina built her own new home nearby. This home sat empty until 1902 and was then occupied by the widow of company president Eugene du Pont. The dwelling was vacant for many years after her passing, and the grounds were donated for the construction of the present-day rehabilitation center. (Above, courtesy of the Winterthur Library: Winterthur Archives; below, courtesy of Hagley Museum and Library.)

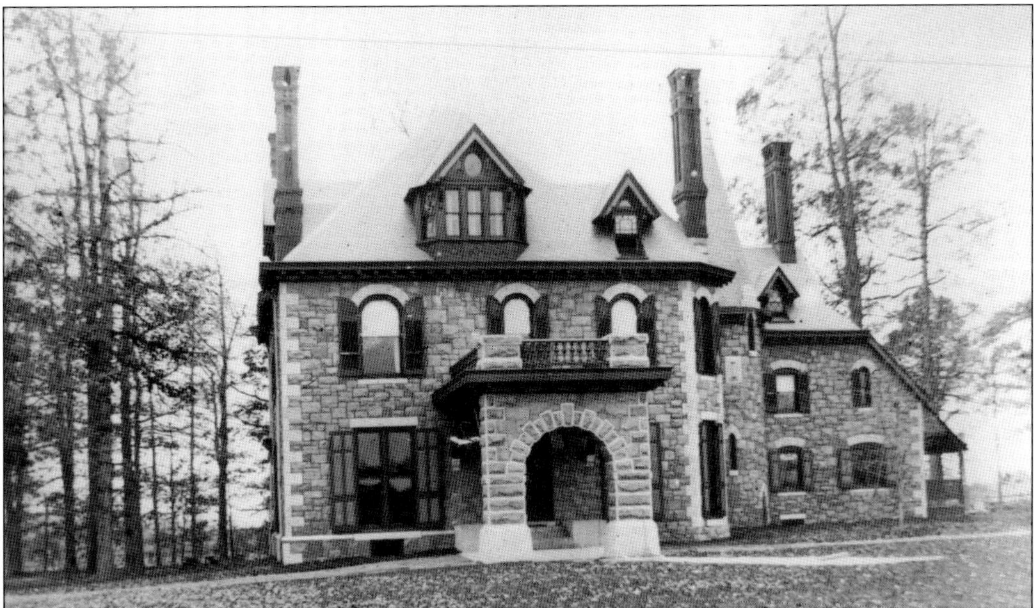

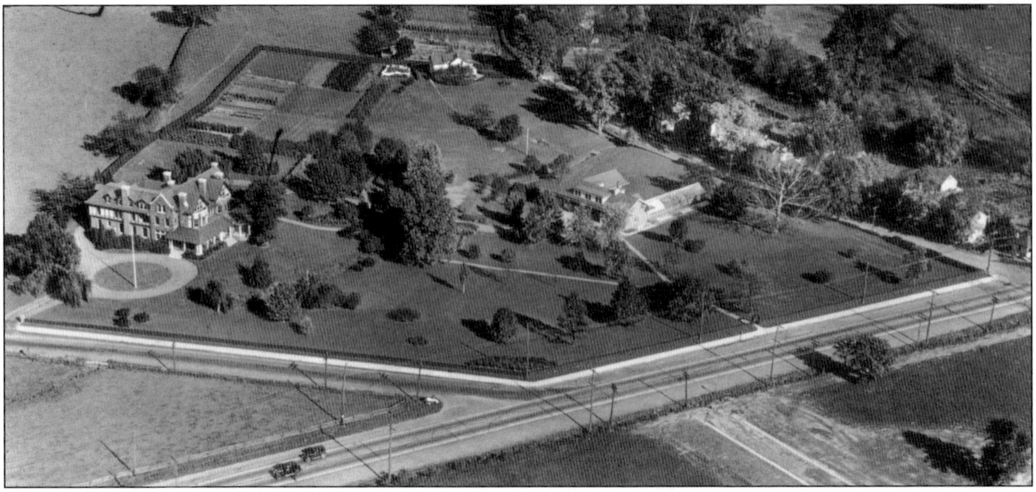

The Dallin Aerial Survey Company captured this view of Irisbrook, the home of William Raskob located at the corner of Kennett Pike and Route 100, in 1927. Kennett Pike runs in the foreground of the image. This residence was later replaced with a more modern dwelling, now home to the Raskob Foundation for Catholic Activities. (Courtesy of Hagley Museum and Library.)

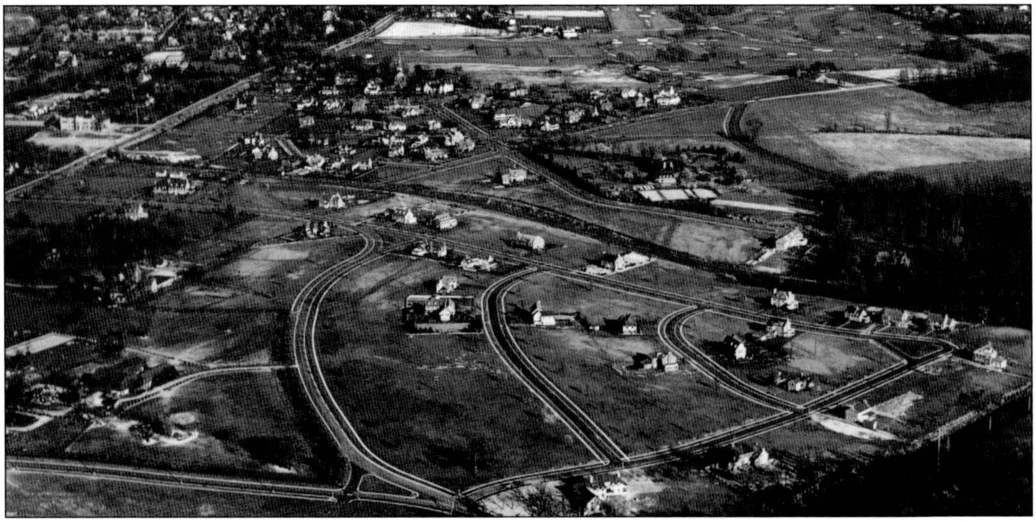

The upscale, suburban neighborhood of Westover Hills was a planned, deed-restricted community developed by William du Pont and his son William du Pont Jr. The community featured large executive homes. The project was begun in 1927 and is shown here in 1935. Kennett Pike runs at the left of the frame, in front of A.I. du Pont School. (Courtesy of Hagley Museum and Library.)

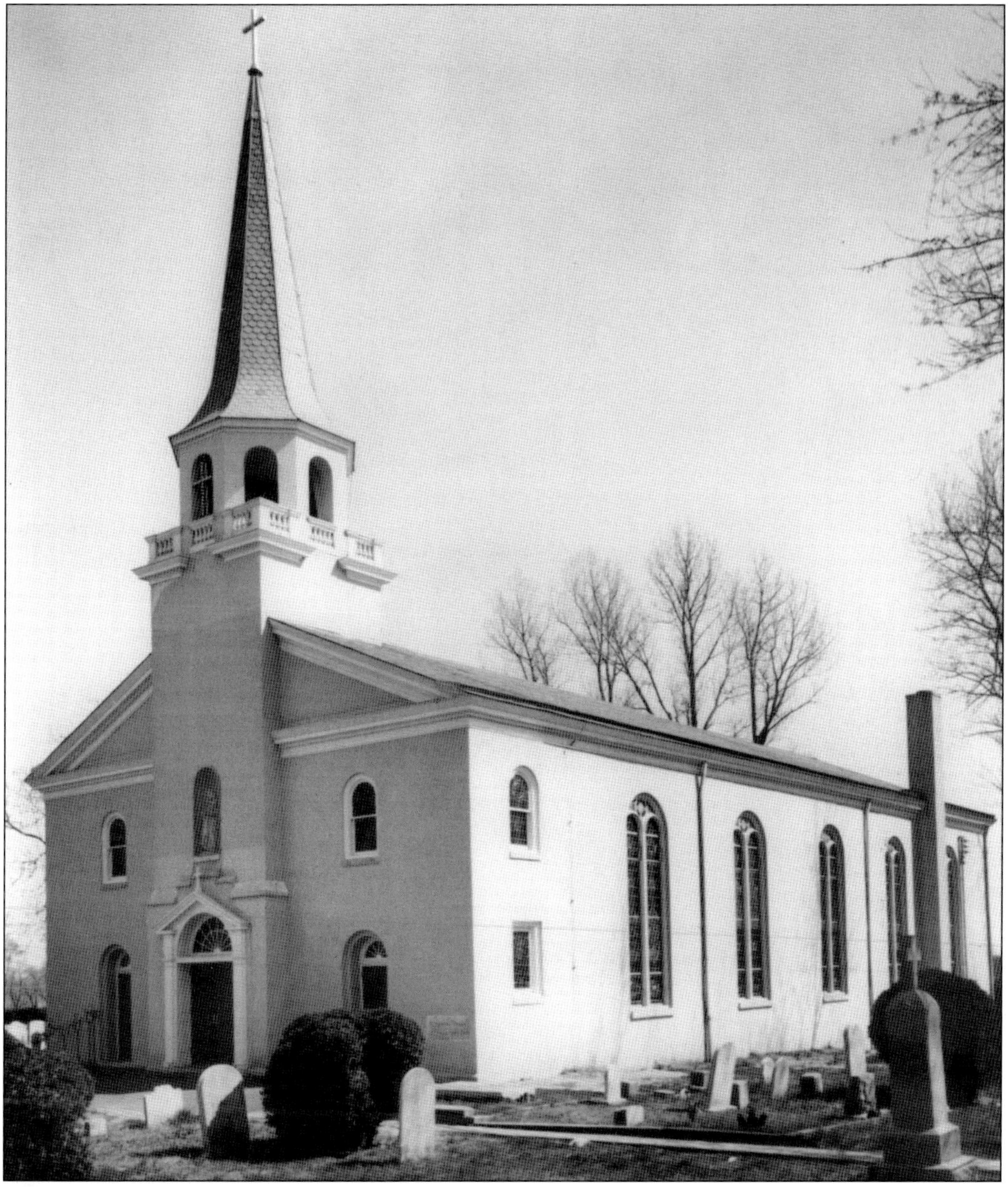

St. Joseph on the Brandywine was built in 1841, largely through the efforts of the DuPont Company. DuPont understood that by providing for their workers in the powder mills (who were largely Irish Catholics), the company would benefit. Interest in the church was garnered by Mrs. Victor du Pont, who hosted Mass in her home, and later, Charles I. du Pont, who donated the land on which the church was built. As the congregation grew, Alfred I. du Pont provided for its electrification, rectory, convent, and school. The church even got its yellow color from Alfred du Pont when he had it painted the same color as his home. By 1894, St. Joseph was the largest Catholic church in the state. Presently, the church retains little of its original form. Additions were made in 1848, 1878, 1941, and 1950, with interior renovations taking place in both 1894 and 1941. It was listed in the National Register of Historic Places in 1976. (Courtesy of Hagley Museum and Library.)

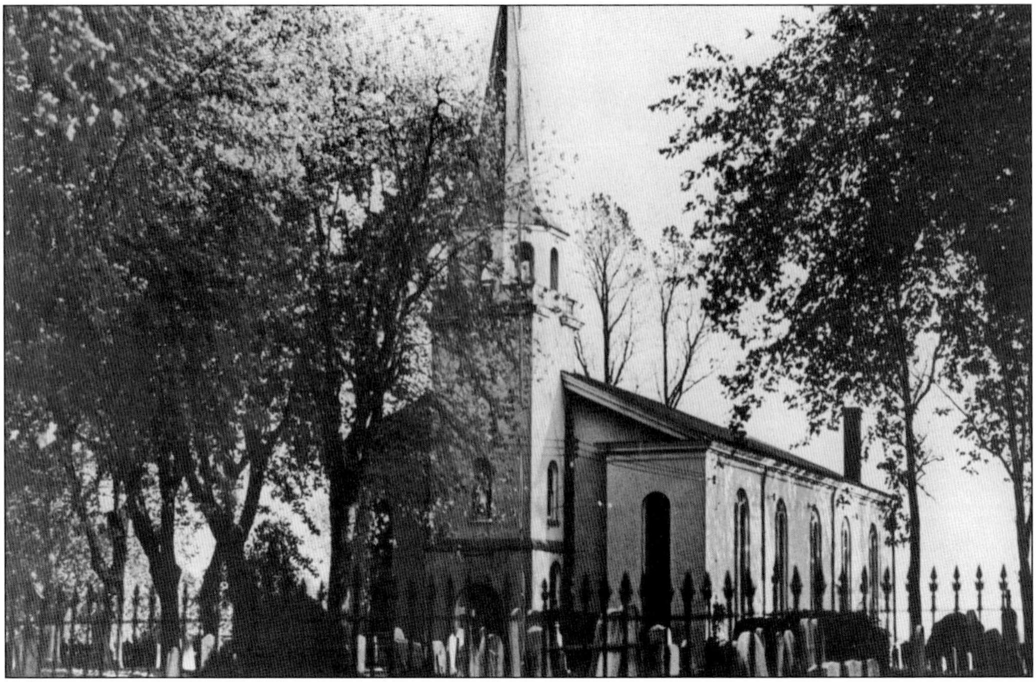

An earlier view of St. Joseph on the Brandywine shows the church between 1930 and 1950. While the powder mills were in operation, pew rental could be deducted directly from Du Pont's payroll. The property is bounded by Kennett Pike and Route 100 (Montchanin Road). (Courtesy of Hagley Museum and Library.)

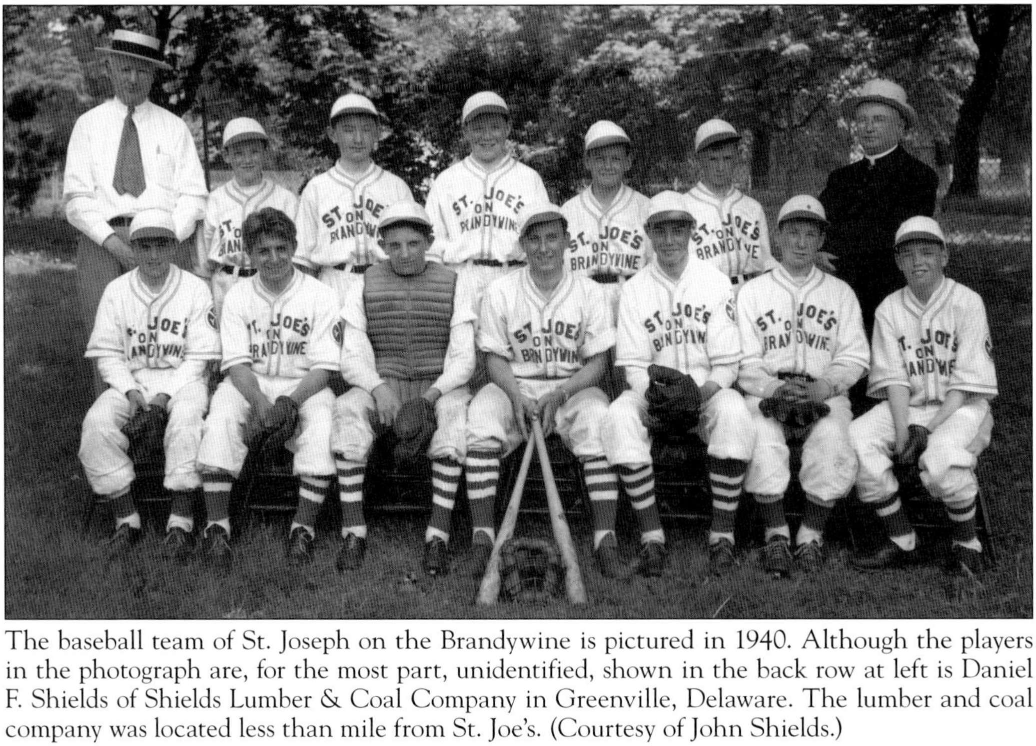

The baseball team of St. Joseph on the Brandywine is pictured in 1940. Although the players in the photograph are, for the most part, unidentified, shown in the back row at left is Daniel F. Shields of Shields Lumber & Coal Company in Greenville, Delaware. The lumber and coal company was located less than mile from St. Joe's. (Courtesy of John Shields.)

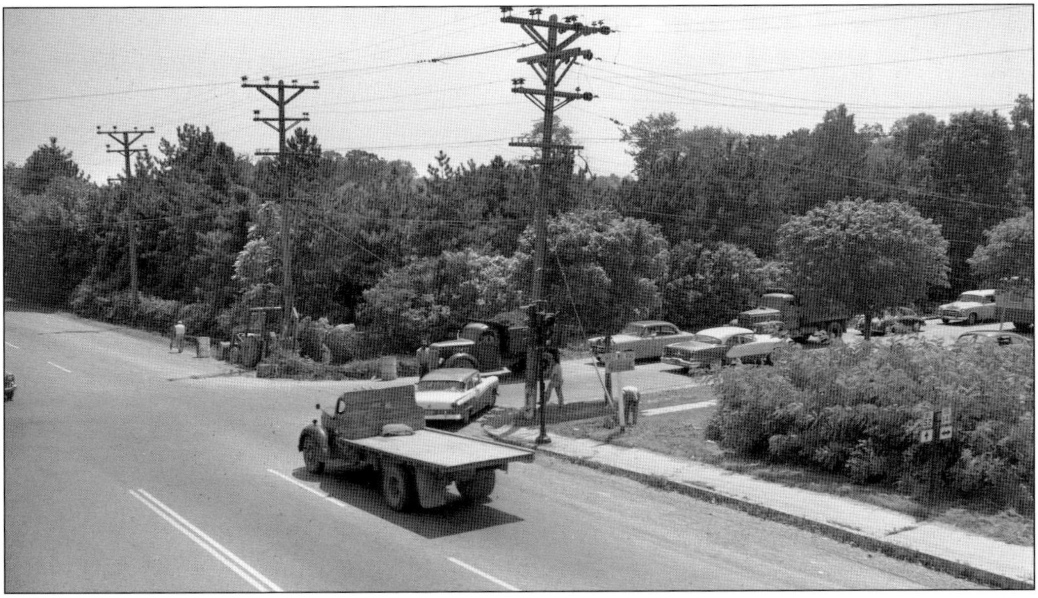

These two August 1958 photographs show the intersection of Kennett Pike and Barley Mill Road (also known as Route 141) before the construction of the overhead crossing of Kennett Pike. Both views are looking toward the city of Wilmington. In the image below, a left turn would direct travelers toward the DuPont Experimental Station, and a right would lead toward Prices Corner. Note the small sign for the Brandywine Raceway at right with the signs marking the routes, as well as the delivery truck for the *News Journal* newspapers. (Both, courtesy of the Delaware Public Archives.)

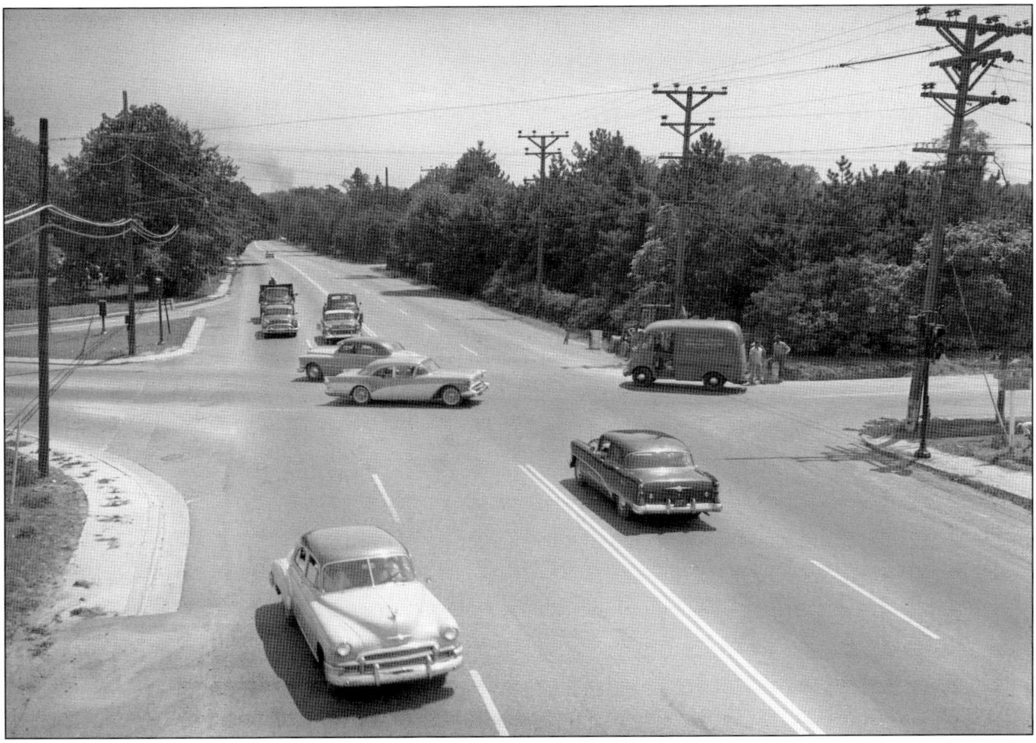

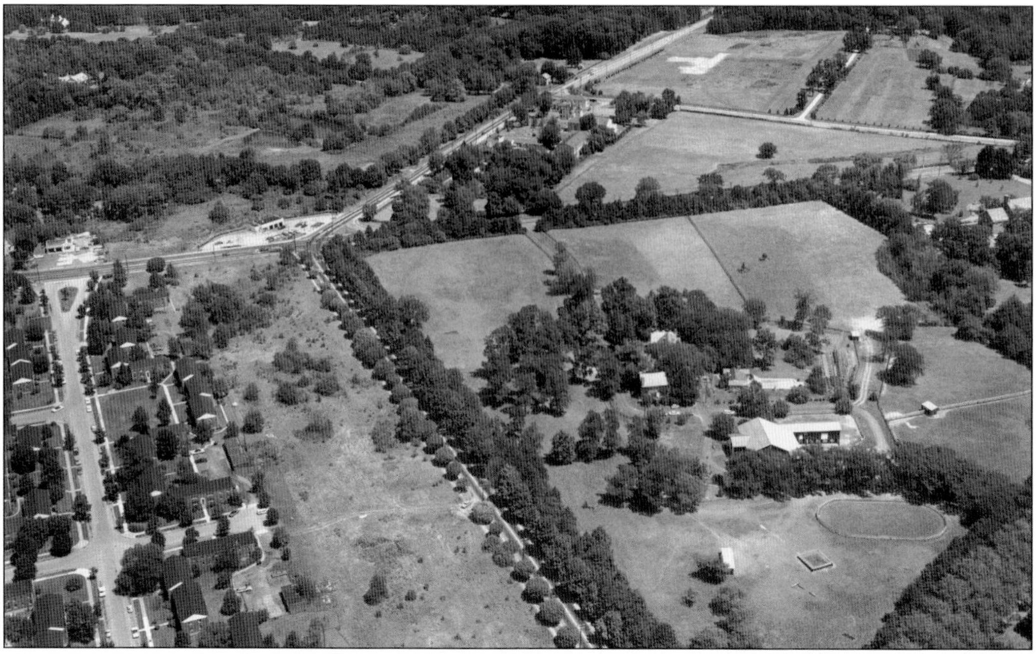

This photograph shows an aerial view of the intersection of Kennett Pike and Route 141 in 1968. At left are the Monroe Park Apartments, presently known as Greenville Place. Kennett Pike runs from right to left in this view. The complex of St. Joseph on the Brandywine can be seen heading away from the pike. (Courtesy of the Delaware Public Archives.)

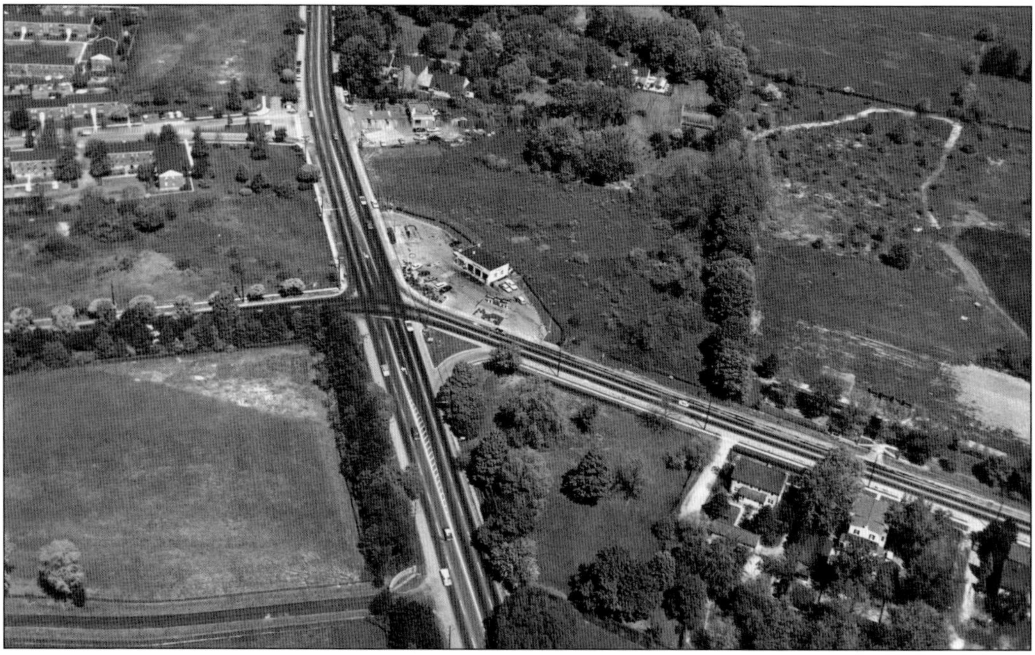

Another aerial view, taken the same year as the photograph above, shows Kennett Pike running through the center of the image. To the north (top of image) is Greenville. Delaware Route 141 runs to the right of the photograph toward the Hagley Museum. (Courtesy of the Delaware Public Archives.)

Two

Greenville

Located just north of Wilmington along Kennett Pike is an area known as Greenville. Today, Greenville is one of the premier addresses in northern New Castle County, dotted with well-to-do suburban neighborhoods and upscale retail shopping.

In its early days, however, Greenville would have been easy to miss driving along Kennett Pike, with only a few du Pont estate gates and a railroad crossing. By the second decade of the 20th century, Shields Lumber & Coal Company dominated the suburban landscape with its retail store, lumberyard, and other buildings. Tucked in amongst these was the Greenville Post Office. The 1950s saw the construction of the Shields Suburban Shops, an early strip mall that included Johnson's Pharmacy and Janssen's Market. Janssen's is still in operation today just south of its original location, now situated at 3801 Kennett Pike in a more modern shopping center with other upscale stores.

Greenville takes its name from the Green family, the original operators of Green & McIntyre Lumber Company (which later became Shields). The family lived across from the site of the lumber company along Kennett Pike. The business was sold to employee Daniel F. Shields in 1929.

Once the construction of the Shields Suburban Shops was complete in 1950, Greenville's retail area began to grow. Today, in addition to the complex that was once home to Shields, three additional upscale shopping centers have been built along Kennett Pike. However, turning from Kennett Pike to one of Greenville's back roads—such as Buck Road, Hillside Road, or Campbell Road—leads to some of the finest real estate in Delaware. Many du Pont family members have called Greenville home since the mid-19th century and still do today. On one section of the former estate of Eugene E. du Pont, just north of the retail area, is Twin Lakes Brewery, a local establishment famous for its microbrews.

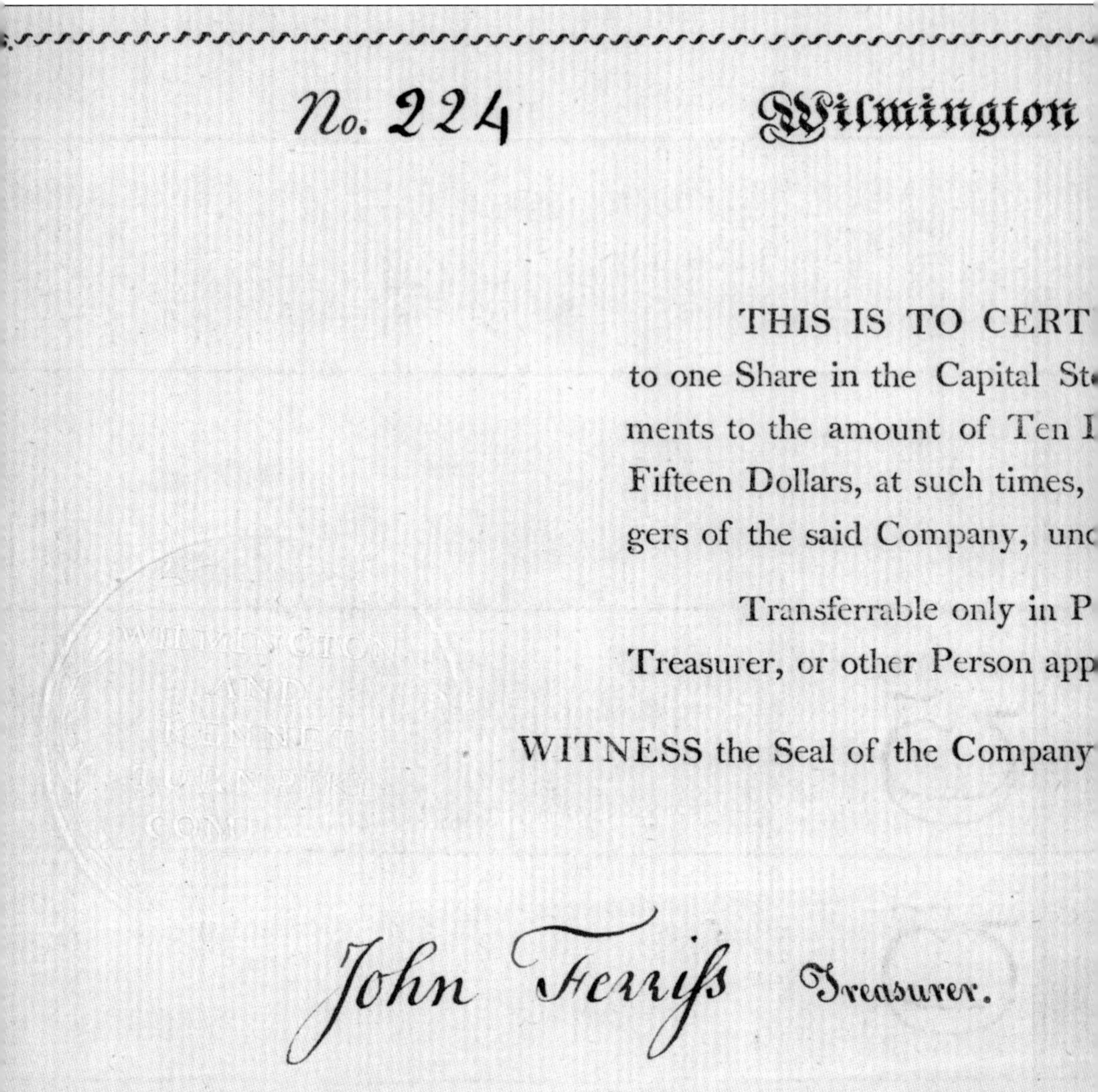

No. 224 Wilmington

THIS IS TO CERT
to one Share in the Capital St
ments to the amount of Ten I
Fifteen Dollars, at such times,
gers of the said Company, unc

Transferrable only in P
Treasurer, or other Person app

WITNESS the Seal of the Company

John Ferris Treasurer.

From the time of its inception, the du Pont family has had strong influence and partial ownership over of the Wilmington & Kennett Turnpike. E.I. du Pont, founder of the company that bears his name, and many of his children were stockholders in the turnpike company that was charted in 1811. Share No. 224 belonged to E.I.'s oldest child, Victorine (1792–1861). Widowed after six months

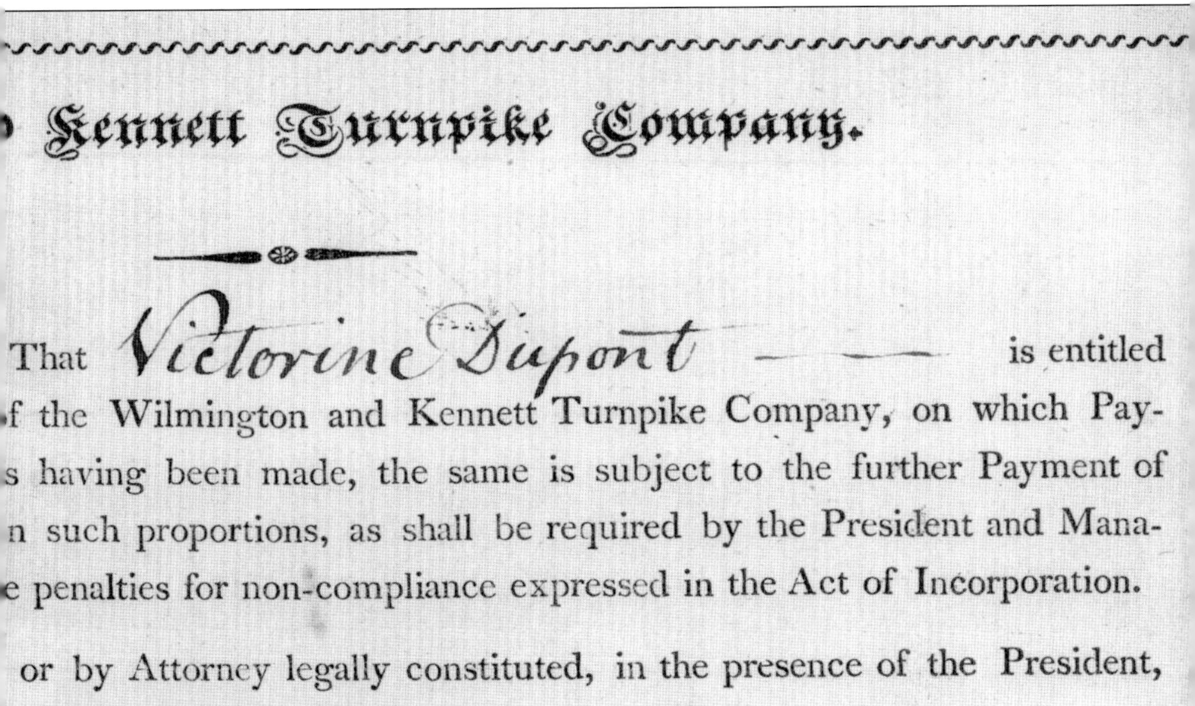

Kennett Turnpike Company.

That Victorine Dupont is entitled
f the Wilmington and Kennett Turnpike Company, on which Pay-
s having been made, the same is subject to the further Payment of
n such proportions, as shall be required by the President and Mana-
e penalties for non-compliance expressed in the Act of Incorporation.

or by Attorney legally constituted, in the presence of the President,
l by the President and Managers to receive the same.

Sixth Day of the 7th month 1811

James Jefferis President.

of marriage, Victorine took over the duties at the family home Eleutherian Mills after her mother was injured following an explosion in the powder mills. She never remarried. Victorine was the superintendent of the Brandywine Manufacturers' Sunday school, and the building still stands on the grounds of the Hagley Museum and Library. (Courtesy of Hagley Museum and Library.)

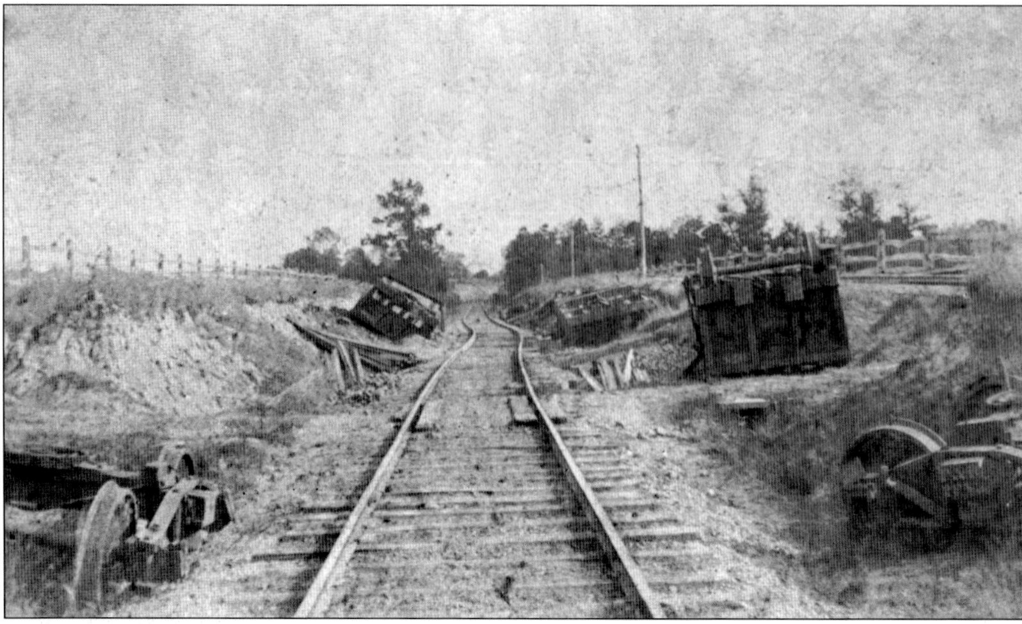

This photograph, dated 1886, shows the debris following a train wreck on the Wilmington & Northern line just west of Kennett Pike behind the present-day Greenville Place apartment complex. In its early days, there was not much in Greenville except for the railroad crossing, a tavern, a local coal and lumber company, and various estates belonging to du Pont family members or DuPont Company executives. (Courtesy of Hagley Museum and Library.)

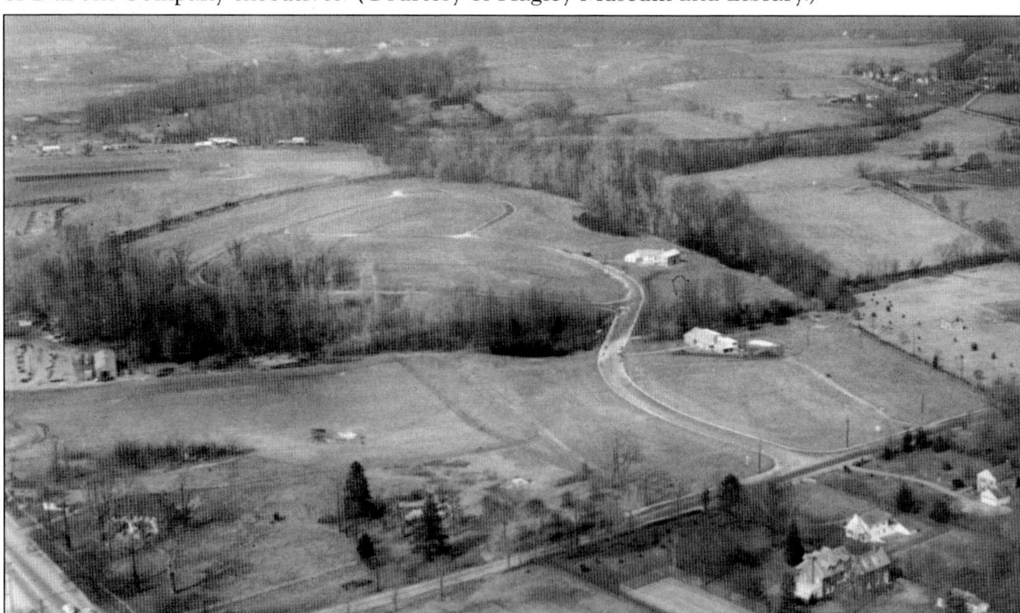

An aerial view of Greenville around 1960 shows the intersection of Kennett Pike (far left) and Buck Road (bottom of frame). The open space at the center of this photograph is the site of the present-day Greenville Center shopping and office complex. Viewers can also see the first few homes being built along Ardleigh Drive, to the east of Kennett Pike. (Courtesy of Hagley Museum and Library.)

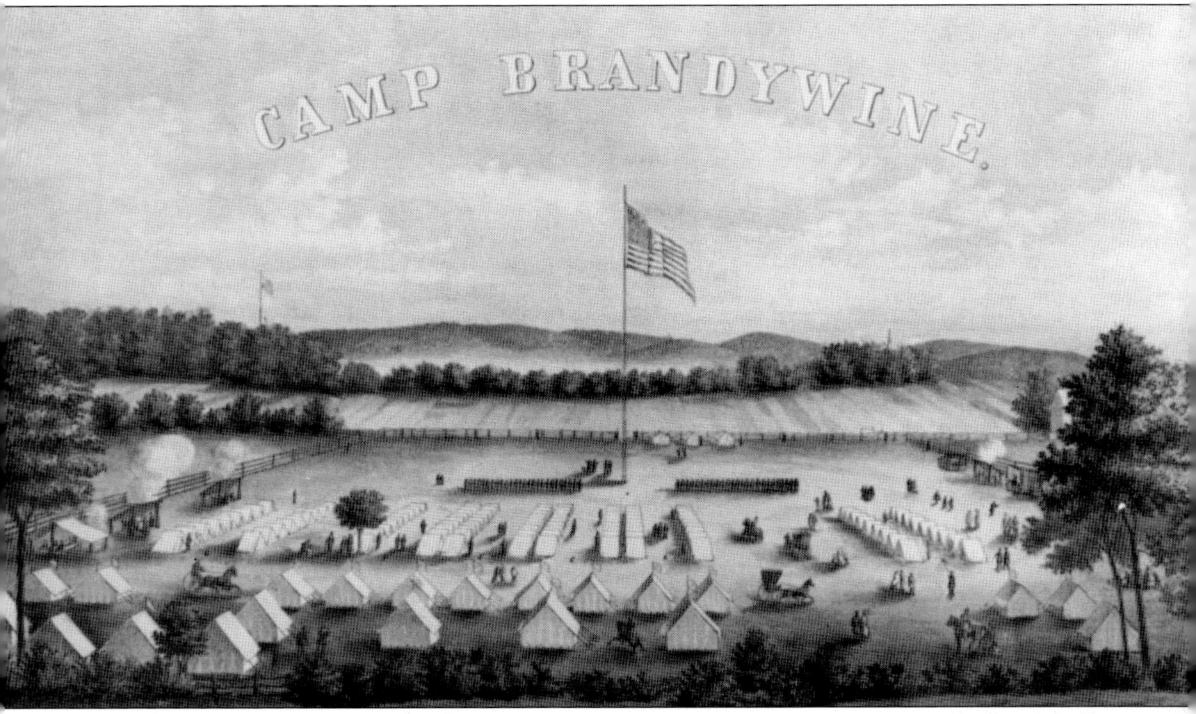

At approximately the site of today's Briars Lane was a camp for Pennsylvania troops that were sent to guard the powder mills during the Civil War in September 1862. The DuPont mills along the Brandywine River were quite important in the supply line of powder to the Union army during the war. The mills were guarded and protected from enemy sabotage. In October 1862, the 4th Delaware Regiment occupied the site. (Courtesy of Hagley Museum and Library.)

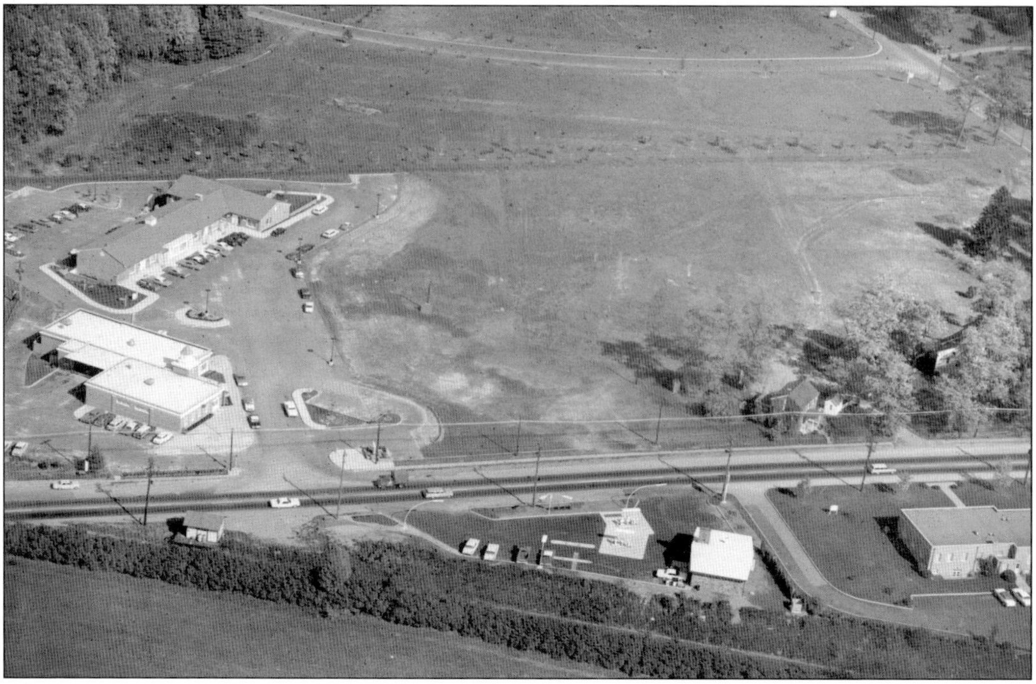

This is another view of Greenville looking east, around 1960, with Kennett Pike running the length of the photograph and Buck Road at the top right of the frame intersecting Ardleigh Drive. Also shown is the Buck Tavern on the same side of the highway as the newly constructed Greenville Post Office. (Courtesy of Hagley Museum and Library.)

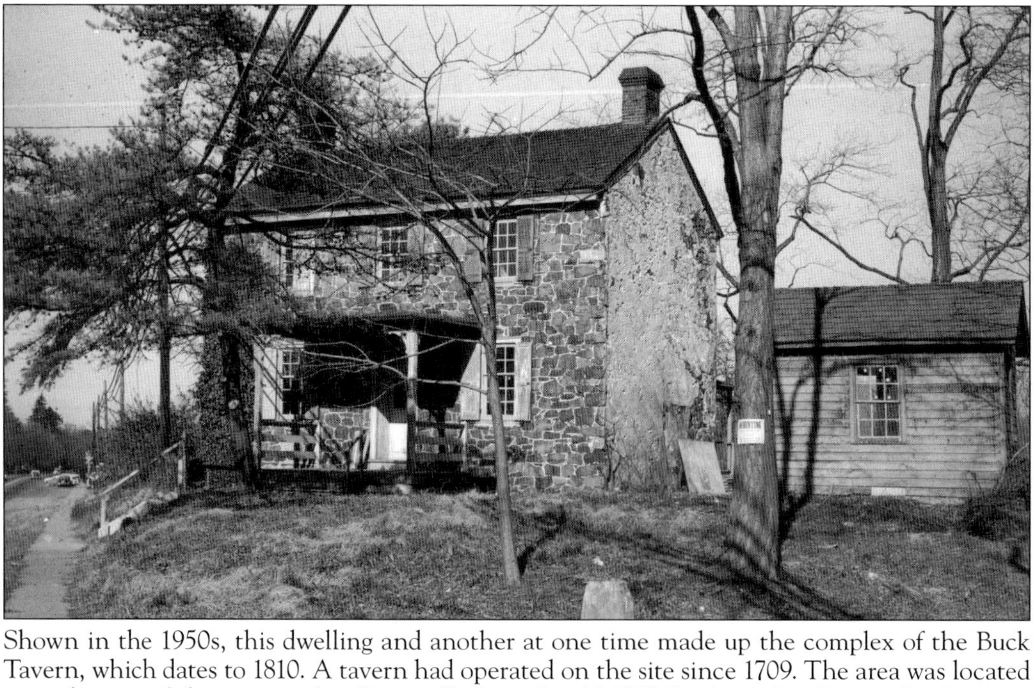

Shown in the 1950s, this dwelling and another at one time made up the complex of the Buck Tavern, which dates to 1810. A tavern had operated on the site since 1709. The area was located near the site of the present-day Greenville branch of M&T Bank in the Greenville Crossing Shopping Center. (Courtesy of Hagley Museum and Library.)

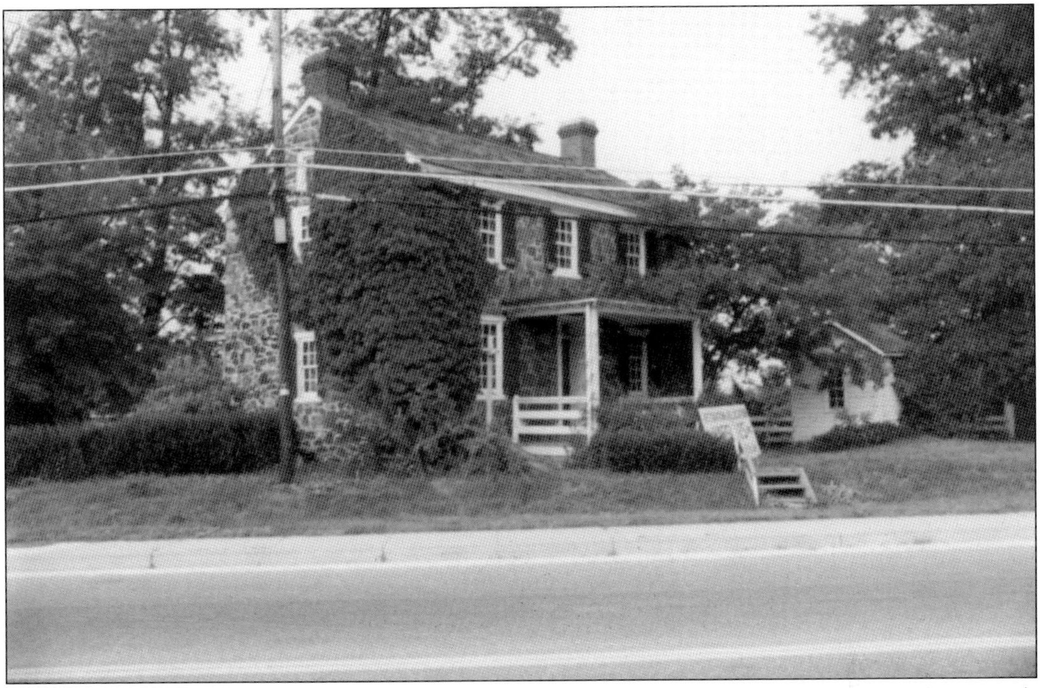

These photographs show the dwelling at the site of the Buck Tavern in the 1970s—above, with Kennett Pike in the foreground; below, with the tavern being moved to its new site along Buck Road around 1975. In 1808, a spy agent for a competitor of DuPont was rumored to be lurking around the tavern. He was confronted and removed from the premises by E.I. du Pont himself. Mr. du Pont was later fined $15 by local law enforcement for disturbing the peace. This dwelling associated with the tavern was moved in 1975 by C. Douglass Buck Jr. and his wife, Jane. (Both, courtesy of George Trapnell.)

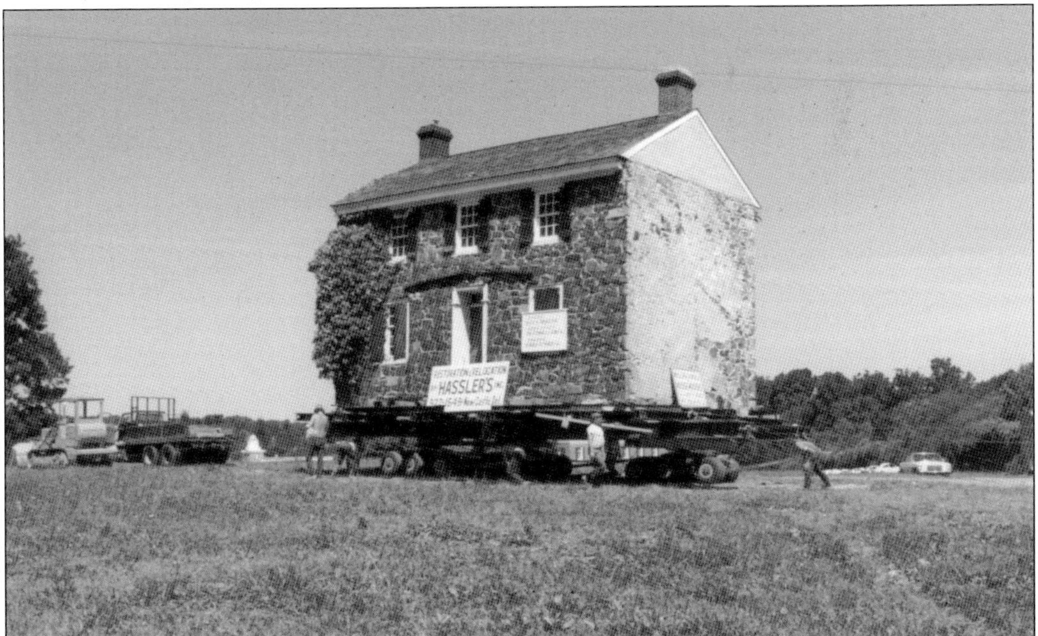

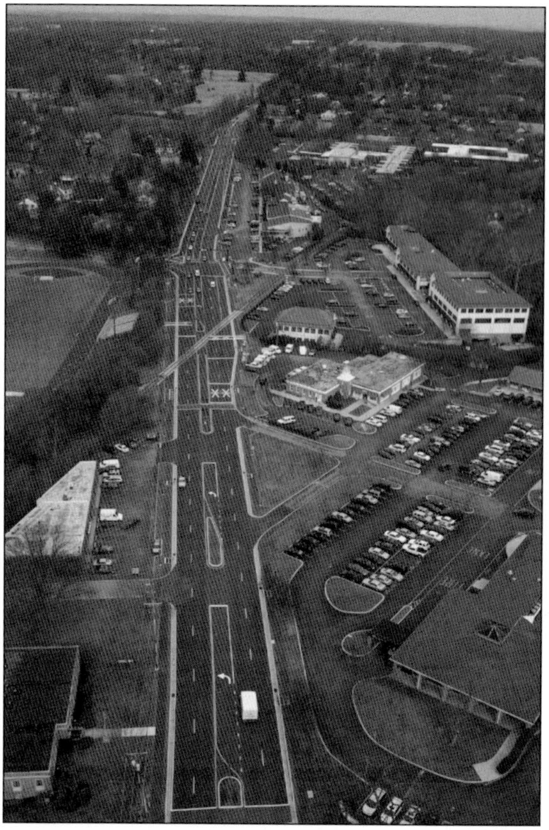

A similar but more modern view of Greenville from the year 2000 is shown here (north is top of the frame). The Greenville Post Office is the building at right with the cupola. Beyond that is Greenville Crossing, the former site of the stockyard of Shields Lumber & Coal Company. At left (just beyond the rail crossing) is the football field and track of Alexis I. du Pont High School (Courtesy of the Delaware Public Archives.)

At the center of this photograph is Dogwood, the estate of Eugene E. and Catherine Moxham du Pont. Mrs. du Pont was afraid of fire, so the entire house was constructed of concrete. It was located on the western side of Kennett Pike, just up Hillside Road, and was demolished in 1966. (Courtesy of Hagley Museum and Library.)

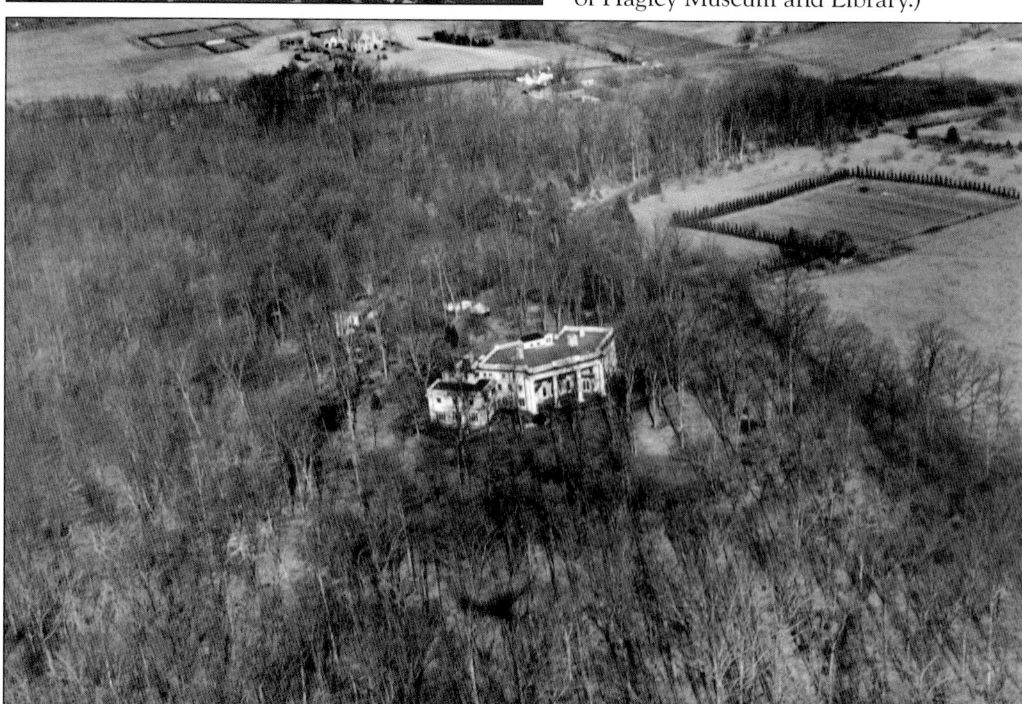

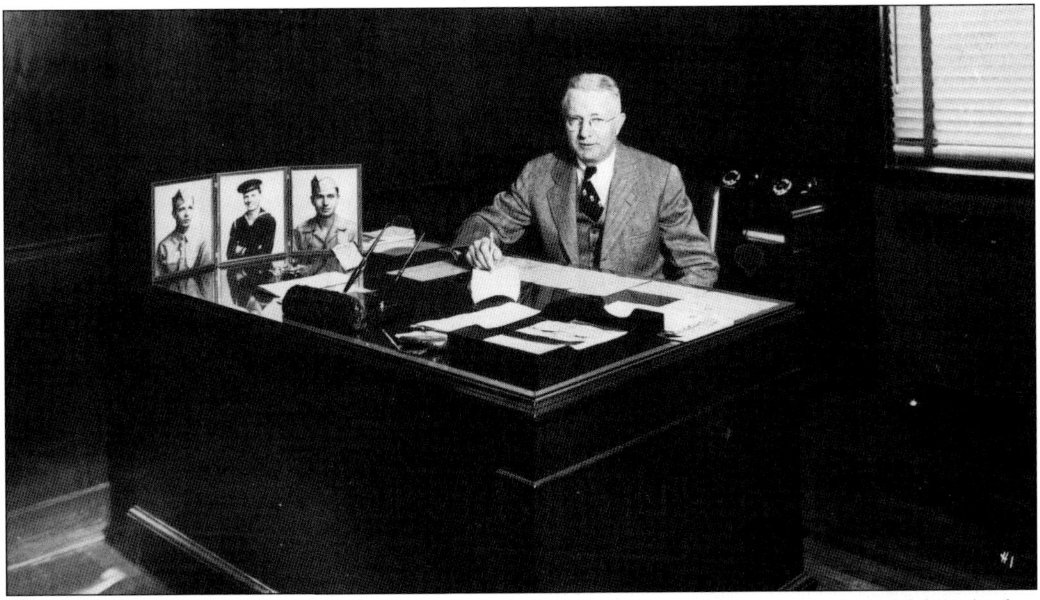

The Shields Lumber & Coal Company was founded in 1929 by Daniel F. Shields Jr. (above) after his purchase of the Green & McIntyre Lumber Company, which had been existence for nearly four decades. A hardware and building-material superstore of its time, its first office was located on the north side of the railroad tracks in Greenville and was demolished in 1940 (below). Within the image of Shields are photographs of his three sons, all of whom served in World War II and eventually became involved in the family business. For many local residents, Shields and Greenville were synonymous for nearly six decades. (Both, courtesy of John Shields and Hagley Museum and Library).

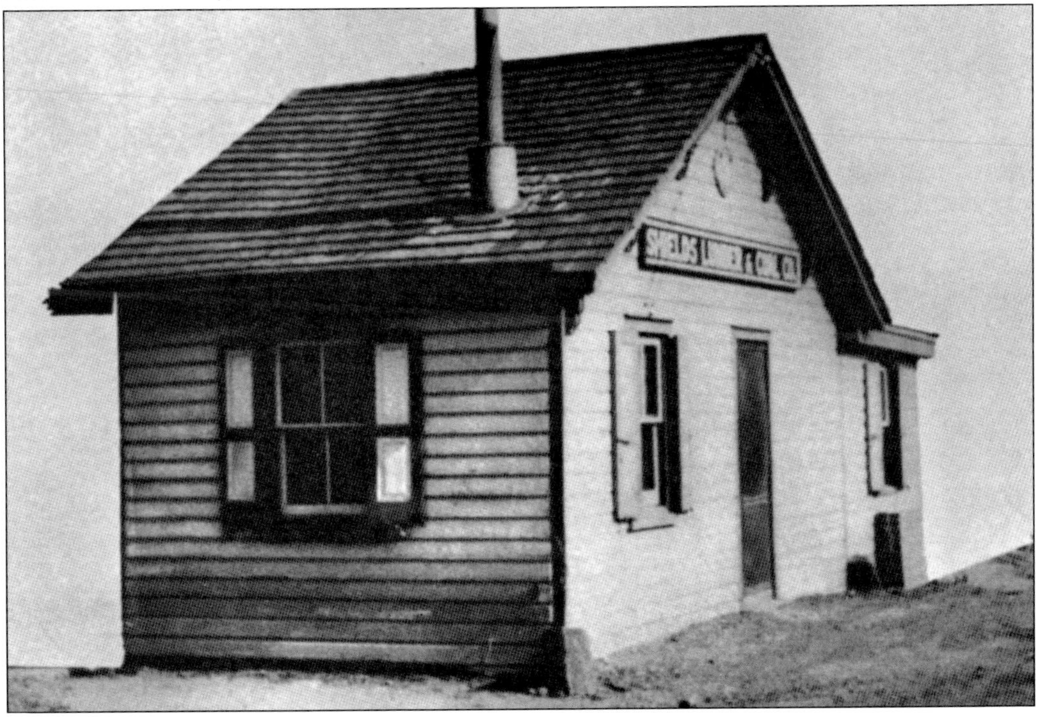

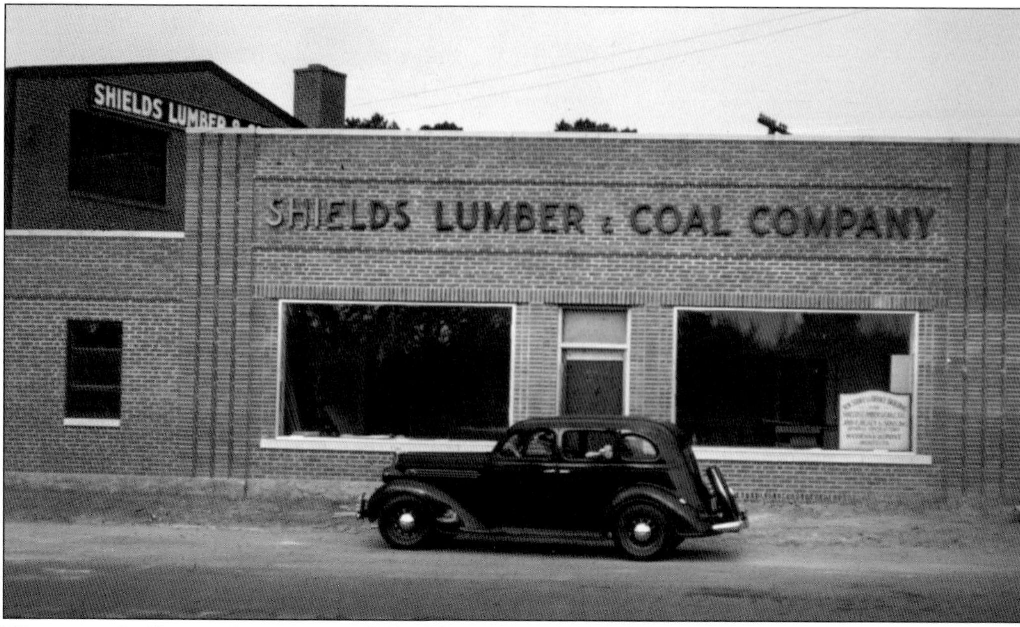

The second office and new store of Shields Lumber & Coal Company was located along Kennett Pike's eastern edge, approximately where the entrance to the Greenville Crossing shopping Center is today. Constructed in 1940, it was designed by Massena & DuPont (who also designed the lower gardens at Alfred I. du Pont's Nemours) and was built by John L. Haley & Sons, Inc. (Courtesy of John Shields.)

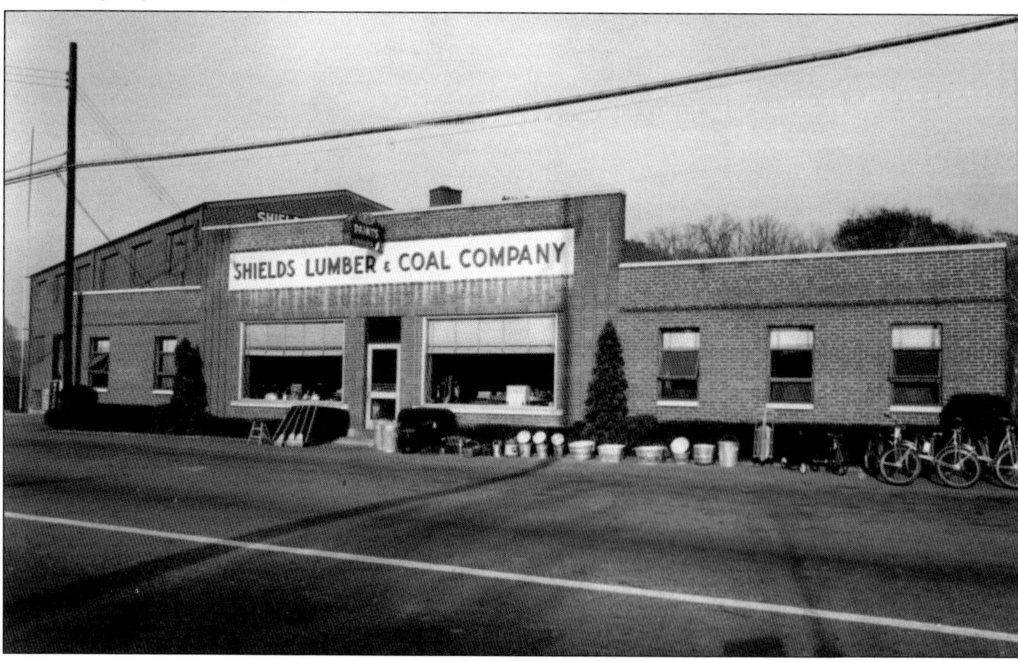

A view of the completed office and store of Shields Lumber & Coal Company around 1947 shows Kennett Pike in the foreground. Shields was a one-stop shop for everything from hardware, tools, building materials, and paint to gardening supplies and bicycles. The warehouse can be seen behind this building. (Courtesy of John Shields.)

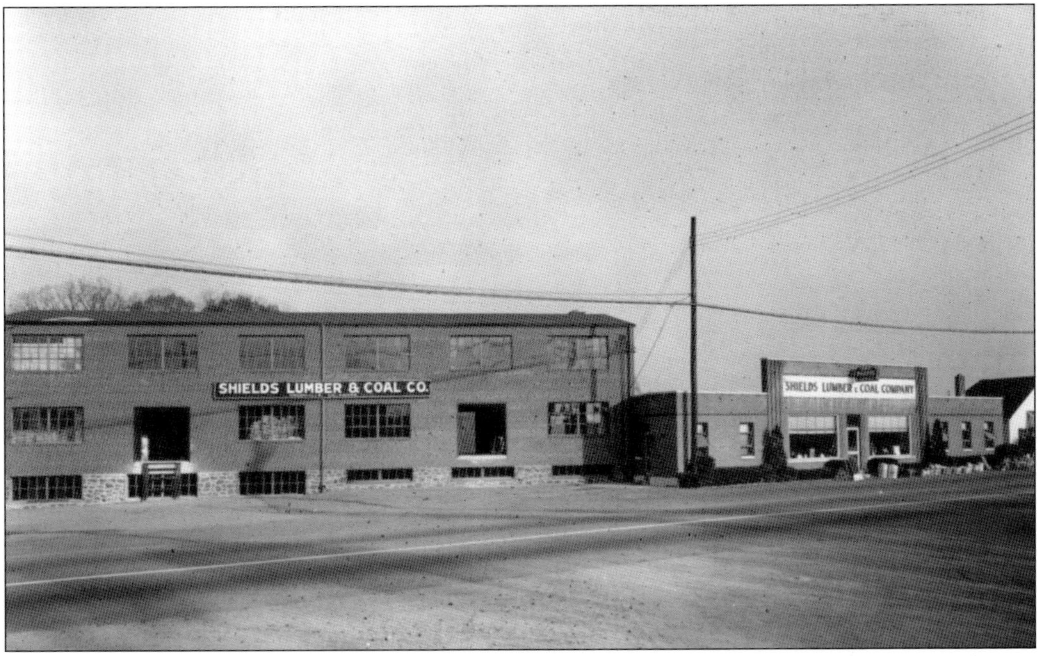

Another view of the Shields complex around the year 1947 shows Kennett Pike, the retail store (right), and the second warehouse of the company, which was constructed in 1936 (left). As of 2013, this is the approximate site of the side parking lot for the Greenville Crossing shopping center. (Courtesy of John Shields.)

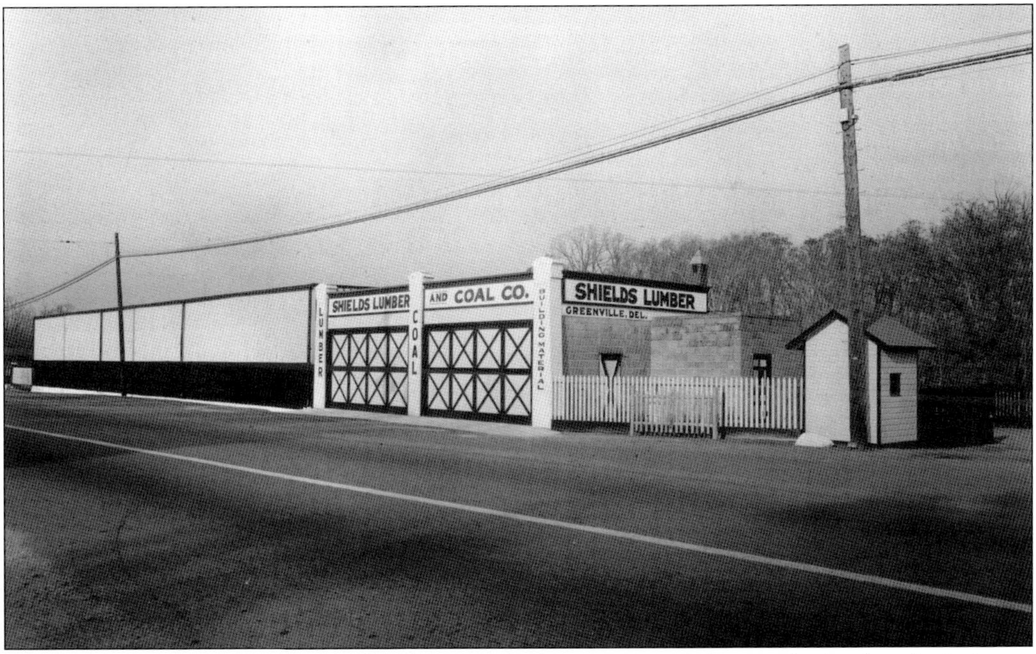

Just past the warehouse and store heading north along Kennett Pike was the lumber shed, shown here in 1947, which was demolished to make way for the Shields Suburban Shops in 1950. Many local estates and contractors went to Shields for building materials in the days before the big-box hardware stores that exist today. (Courtesy of John Shields.)

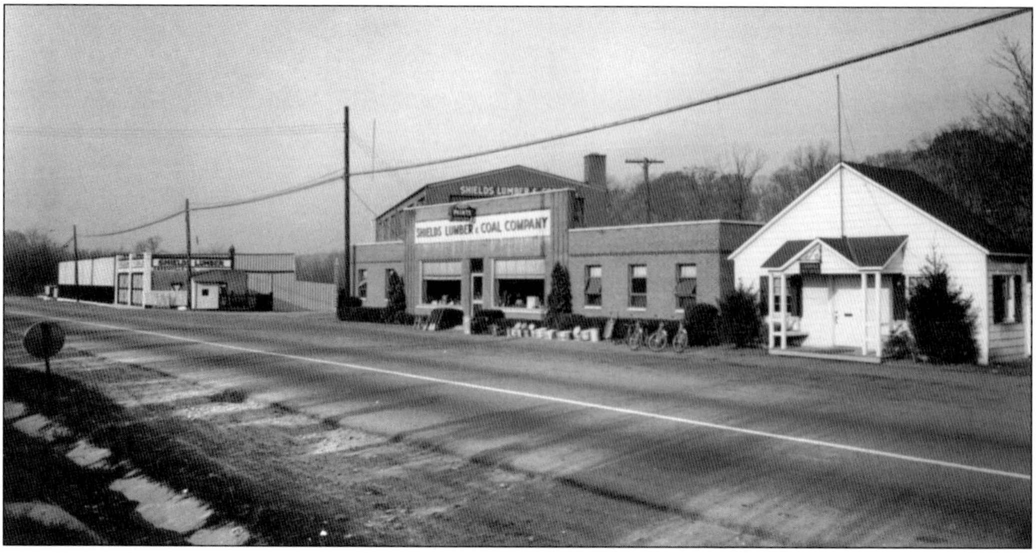

This photograph shows the view of the entire block of Shields Lumber & Coal Company along Kennett Pike in 1947. This view also shows the Greenville Post Office, which was located in the small white building at the far right. It is now located at 5105 Kennett Pike in Greenville Crossing. (Courtesy of John Shields.)

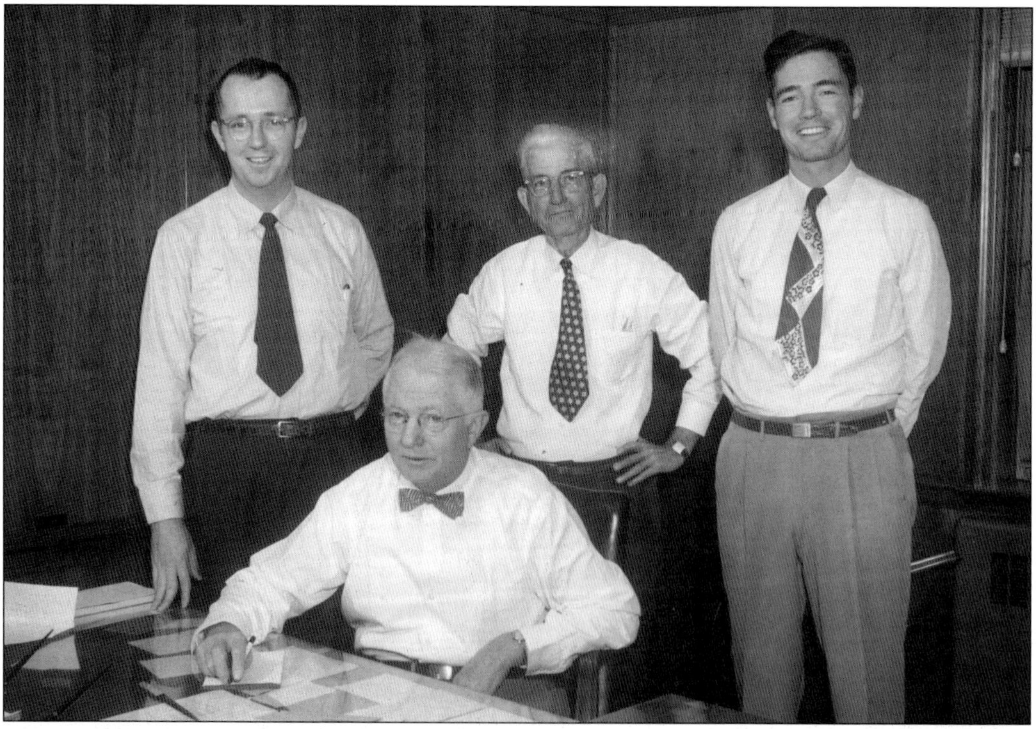

This 1947 photograph shows the president of Shields Lumber & Coal Company, Daniel F. Shields Jr. (seated), and two of his three sons. Standing are, from left to right, John J. Shields, secretary and general manager; Arlie Blevins, longtime Shields employee; and William A. Shields, treasurer. Daniel Shields III became the president of the company following his father's retirement. (Courtesy of John Shields.)

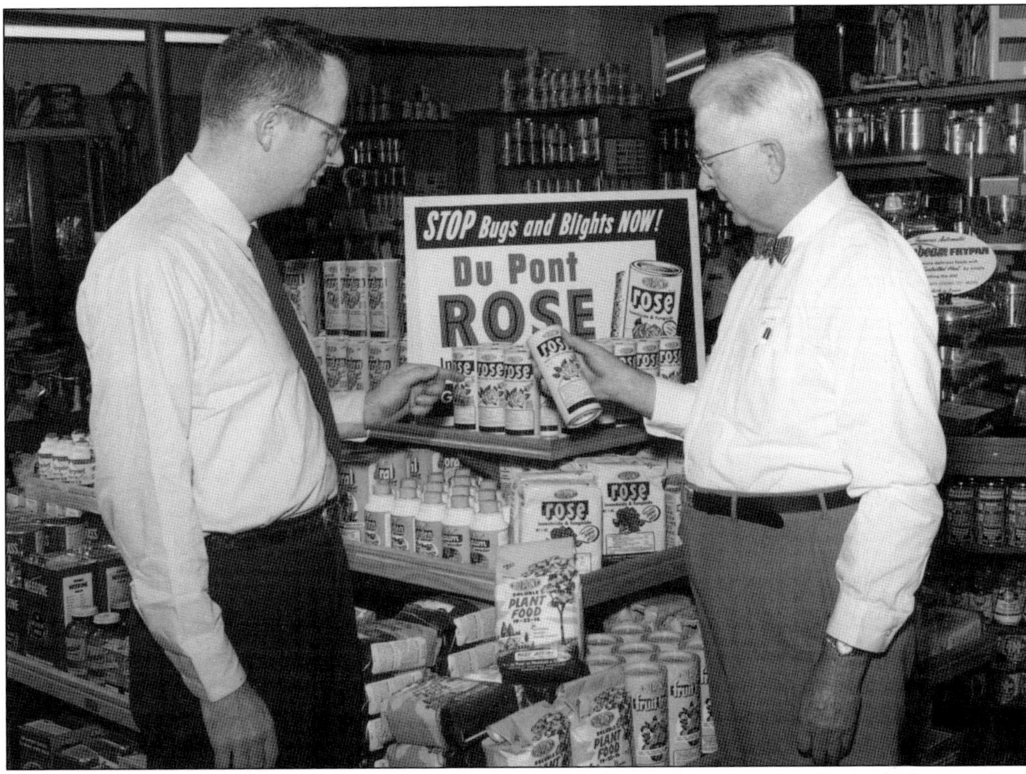

Each of Daniel Shields's sons entered the company in different departments. Daniel III started out in lumber; John, in hardware; and William (Bill), in bookkeeping—all between 1948 and 1955. They also became officers in the Shields Development Company. John Shields (left) is pictured with his father in 1947. (Courtesy of John Shields.)

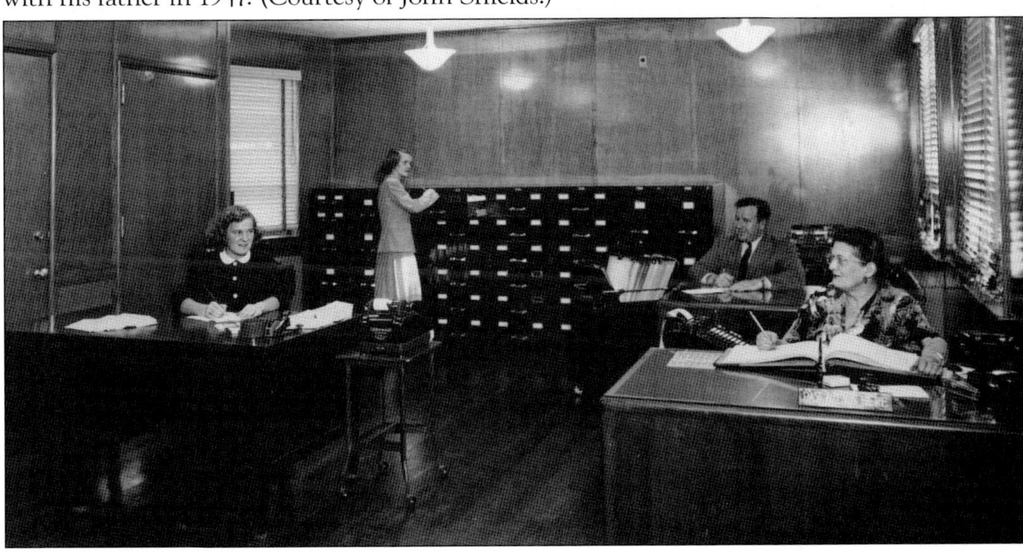

The bookkeeping and accounting department at Shields is pictured in 1947. From left to right are Delores McMahon, Ann Escherich, Walter J. Brady, and Phoebe Sharpley. Brady and Escherich were still at the company as of 1967, the 37th anniversary of the business. (Courtesy of John Shields.)

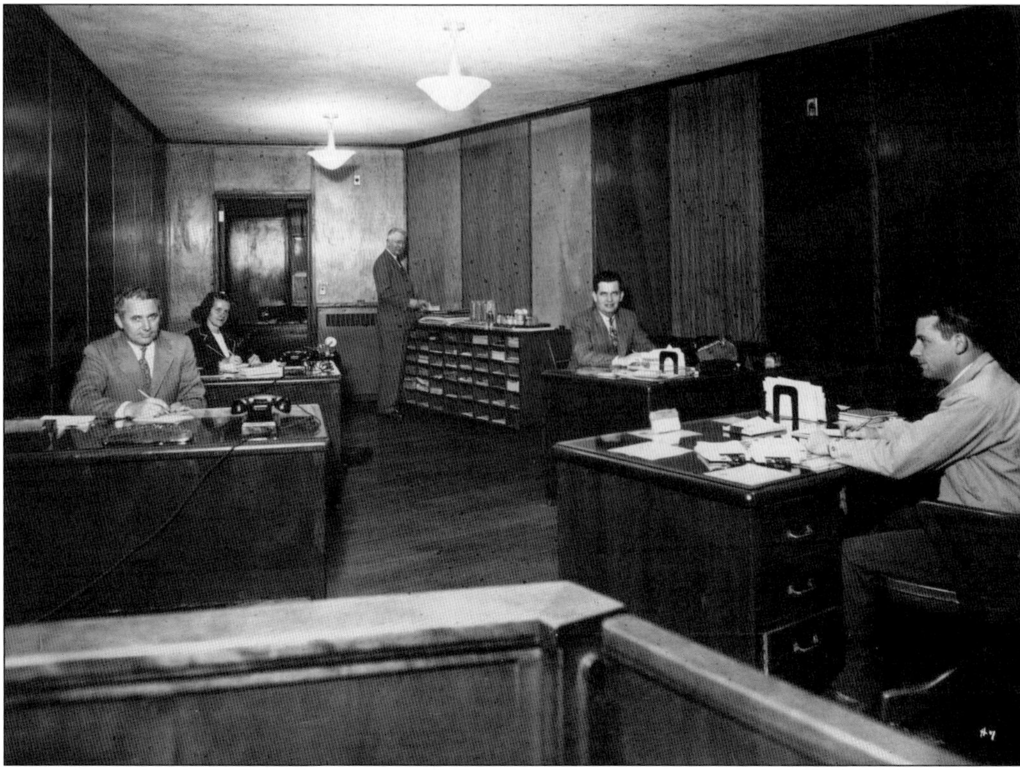

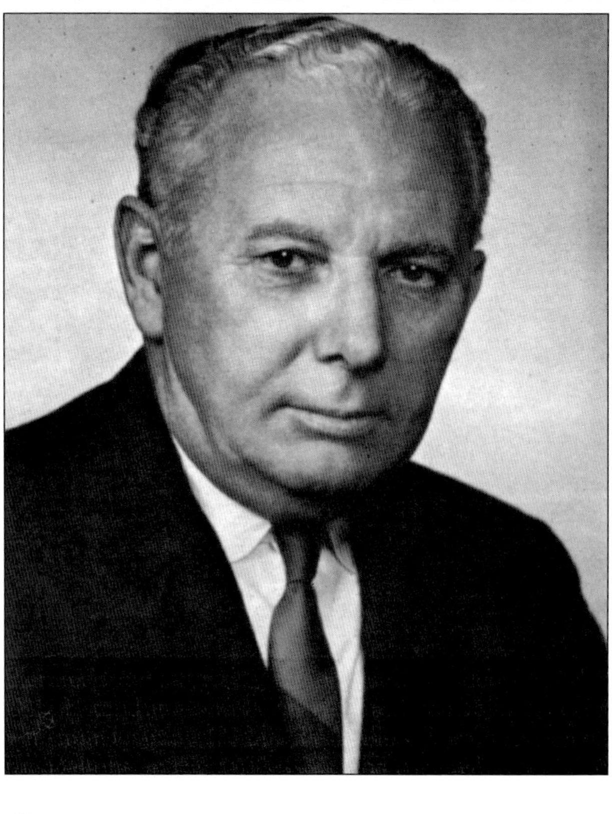

Shown above is the sales department at Shields in 1947. Standing in rear is Daniel F. Shields Jr., and seated are, from left to right, Thomas A. O'Connor, Ann Escherich, John Dougherty, and William Breslin, all longtime employees of the company. Thomas O'Connor, also featured at left, was perhaps the person employed the longest by Shields. He started in 1929 in the days when a salesman also worked in the yard and drove the delivery truck. The Shields family was quoted as saying that O'Connor "came with the place." Generations of the Simpson family also were employed at the store; Eddie Simpson drove a coal truck, Charlie was the manager of hardware stock, and Charles Jr. was a salesman. (Both, courtesy of John Shields.)

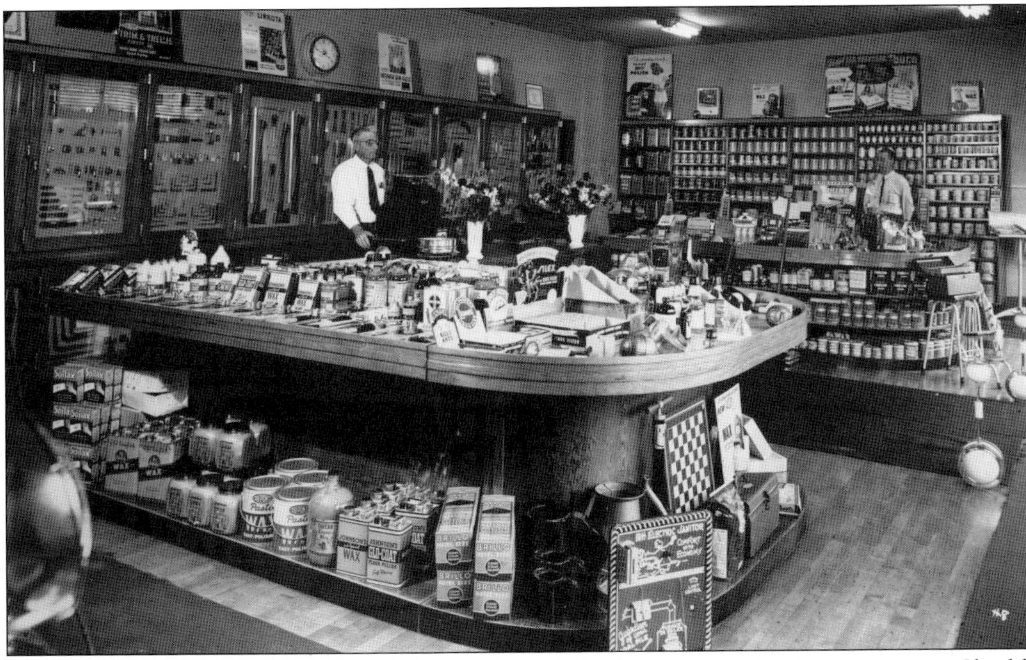

This photograph shows the inside of the retail store's hardware and paint departments at Shields in Greenville in 1947. Charles Simpson Sr. is shown at right, and Eugene Shields is behind the cash register. Note the wide variety of DuPont products carried by the firm. (Courtesy of John Shields.)

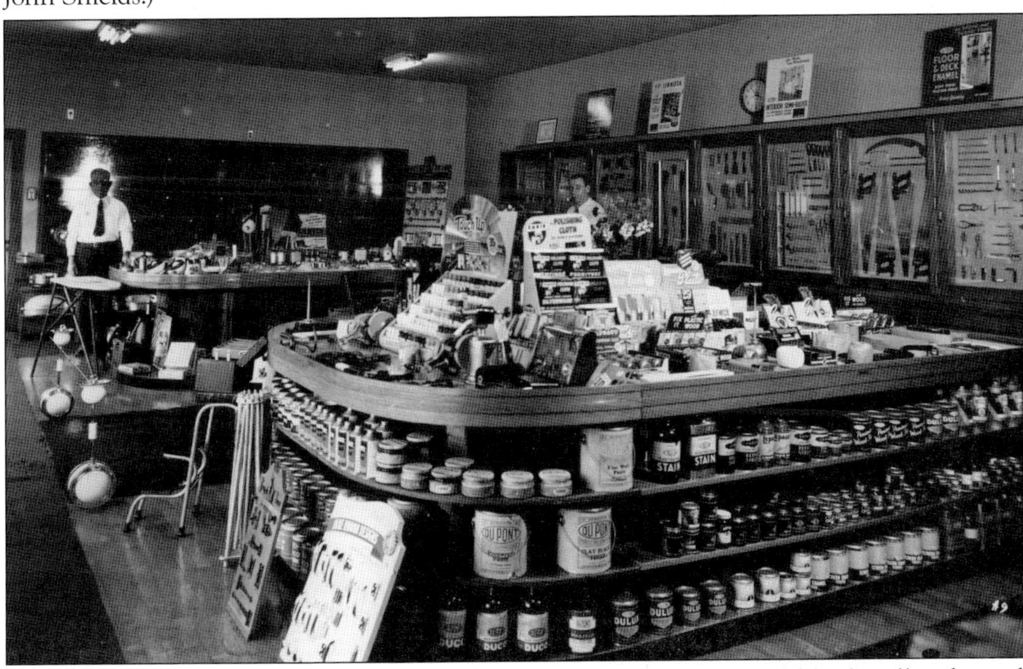

Another view of the store in 1947 shows the sample cases of tools along the back wall and sample towel bars toward the back corner. Sample boards featuring hardware are shown along the floor in the foreground. The store carried a full line of DuPont consumer products. (Courtesy of John Shields.)

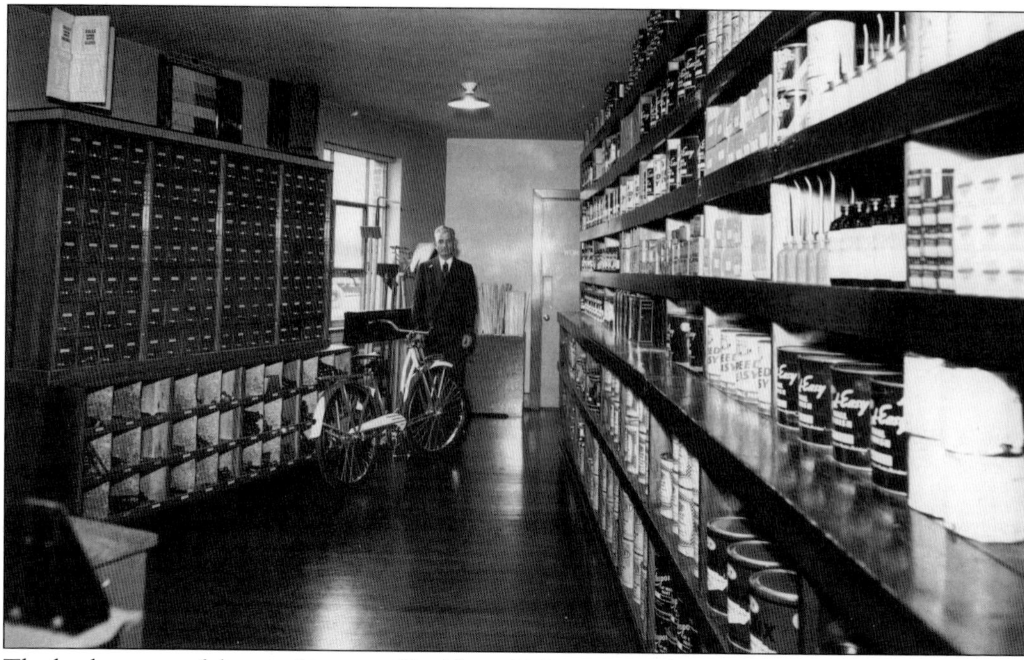

The back corner of the retail store at Shields in 1947 contained an even bigger selection of paints as well as a case of various nuts and bolts (at left). The store also included separate restrooms for women and men; the door to the women's restroom is in the background. (Courtesy of John Shields.)

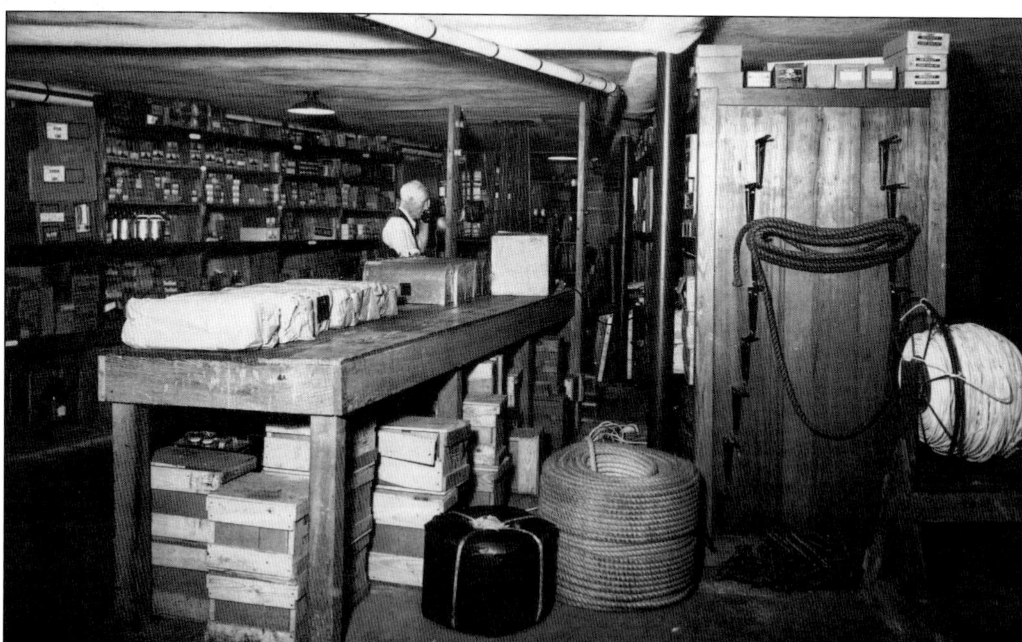

This photograph shows the cellar of the store with employee Philip Dougherty placing hardware and other stock in appropriate storage locations in 1947. Shields was well known for their large stock of items to accommodate every need and project, which was one of the many reasons it was such a success for so many years. (Courtesy of John Shields.)

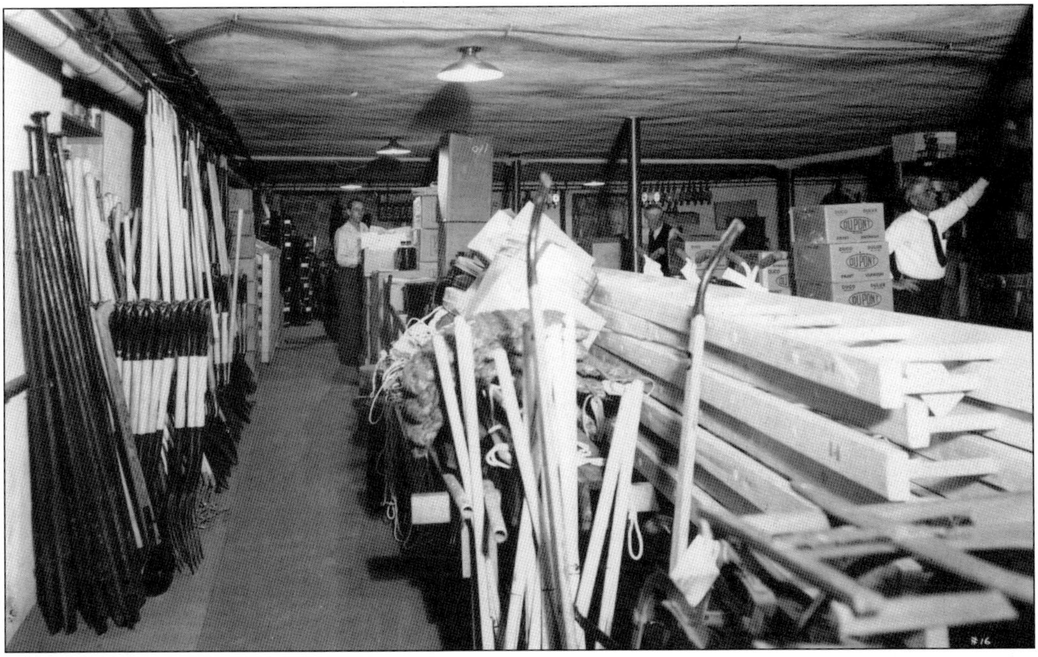

Another view of the cellar in the same year as the photograph above shows the arrangement of hardware and garden supplies. Not only was Shields a building-supply company, it was also a one-stop shop for landscaping tools and equipment. The employees in this photograph are unidentified. (Courtesy of John Shields.)

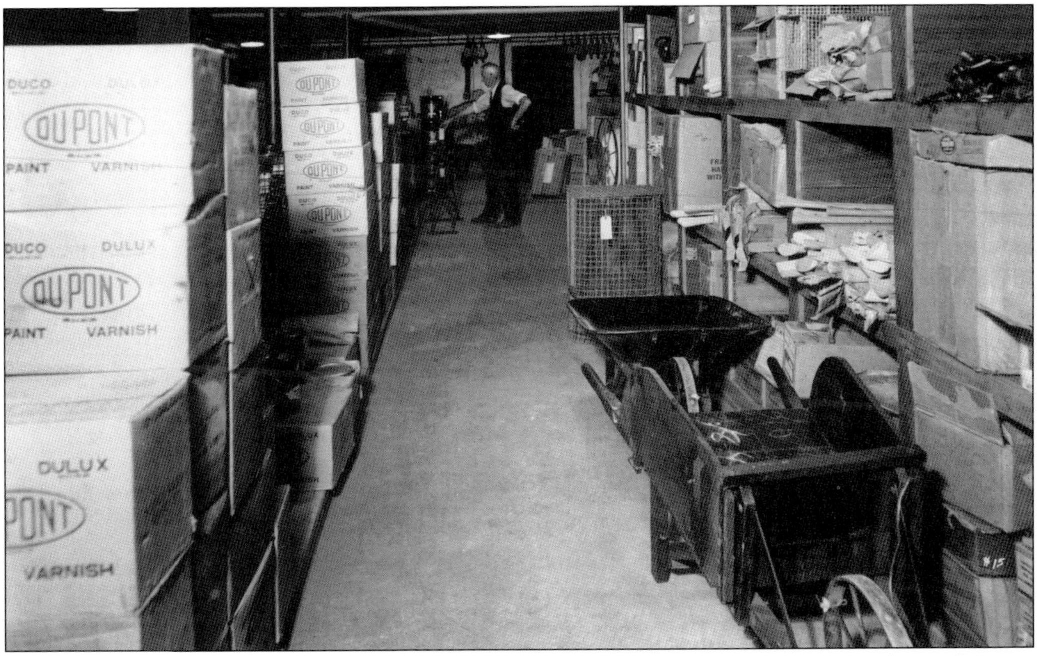

The paint-storage area of Shields was also in the cellar of the office and store, shown here in 1946. Also visible in this photograph are garden supplies, such as wheelbarrows and other items. Employees kept the storage areas orderly to make it easy to retrieve items for customers. (Courtesy of John Shields.)

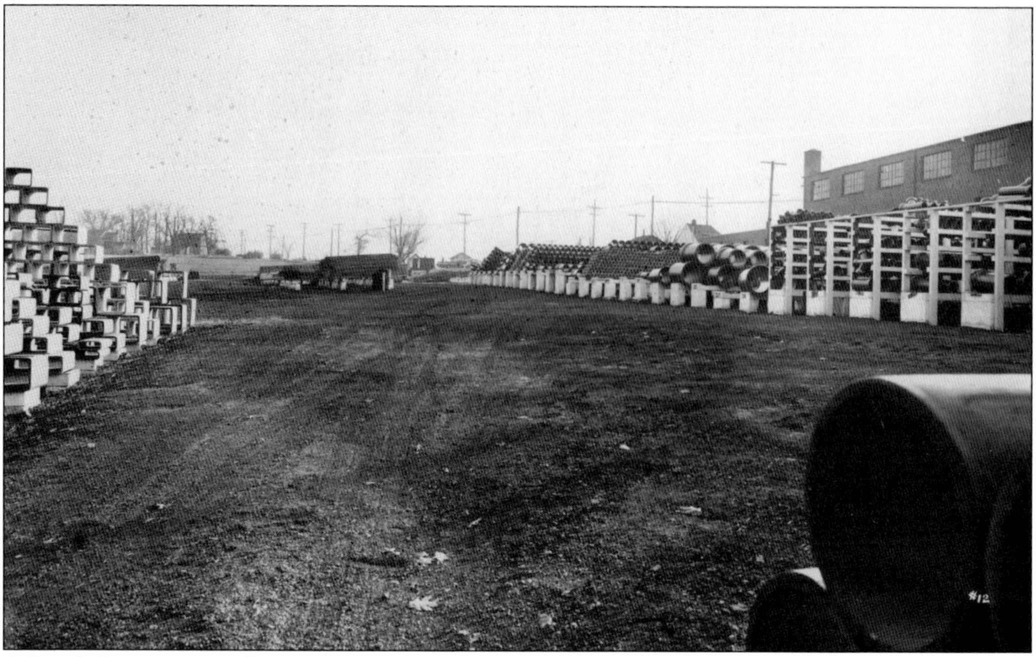

Behind the main warehouse was storage for chimney flues and terra-cotta piping, marked and stored by diameter. This 1946 photograph is looking south, toward Wilmington; note the telephone poles that are running along Kennett Pike. The house and buildings in the background are part of the Buck Tavern complex. (Courtesy of John Shields.)

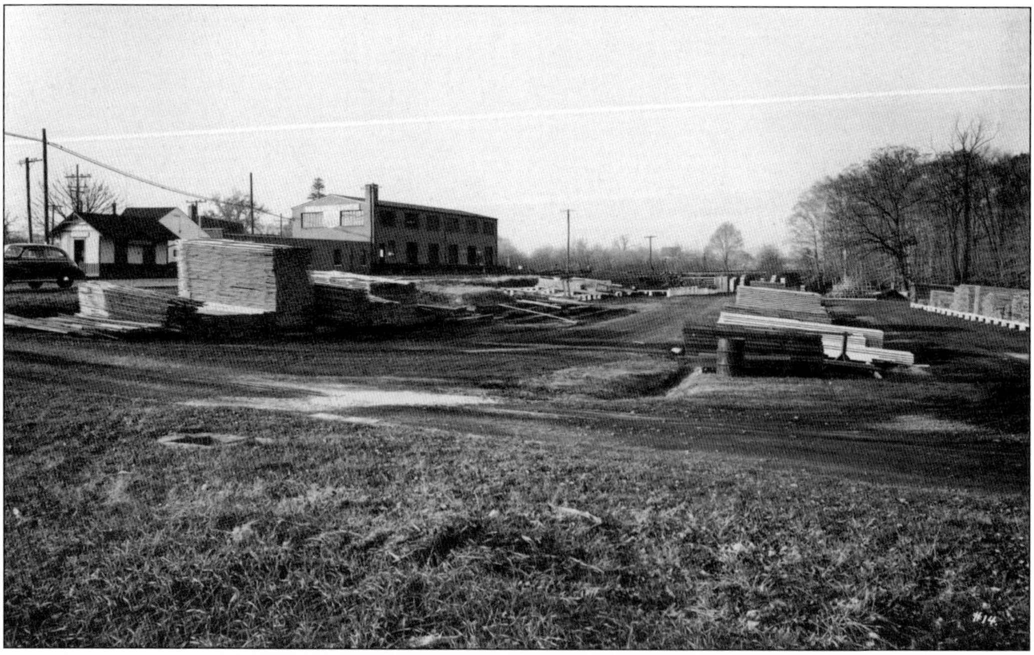

An alternate view of the storage yard at Shields Lumber & Coal Company in 1947 shows piles of Douglas fir, hemlock, and yellow pine timbers. This time, looking north, the railroad tracks are visible at left along with a small railroad station marked "Greenville." Today, this is the location of parking for the Greenville Crossing retail complex. (Courtesy of John Shields.)

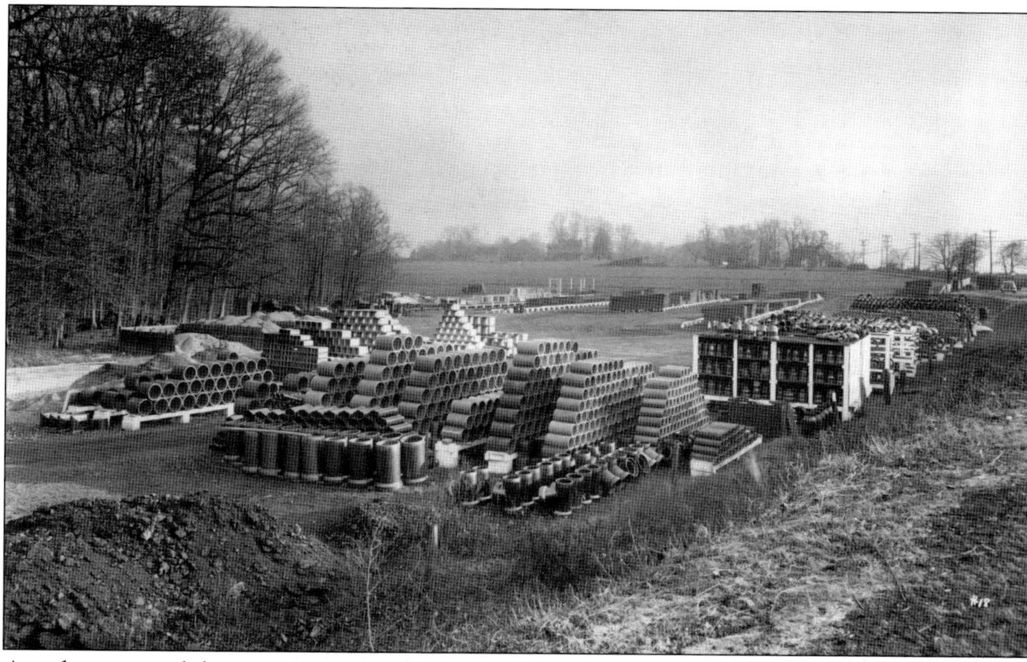

Another view of the storage yard at Shields looks south toward Wilmington in 1947. Kennett Pike is visible at the far right of the frame. The open field just beyond the yard is the site today of the Greenville Center retail and shopping complex, which was begun in the early 1960s by various du Pont family members. (Courtesy of John Shields.)

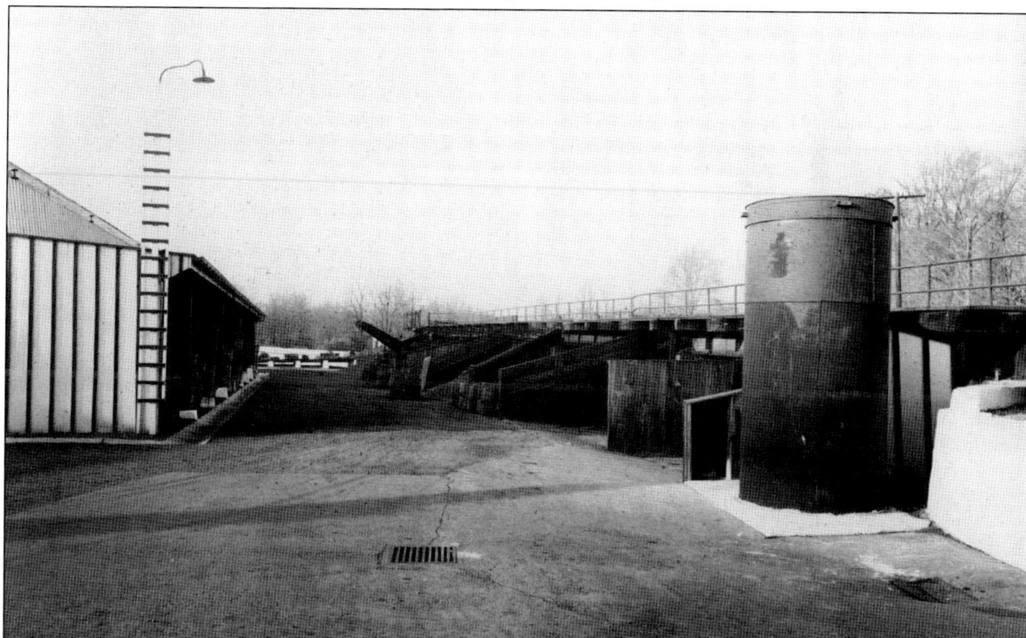

This is a view from behind the lumber shed of Shields that was just past the store and warehouse (heading north) along Kennett Pike. Coal-storage bins are shown at right, and lumber, at left. Shields also made home deliveries of their products, such as coal, if requested by the customer. (Courtesy of John Shields.)

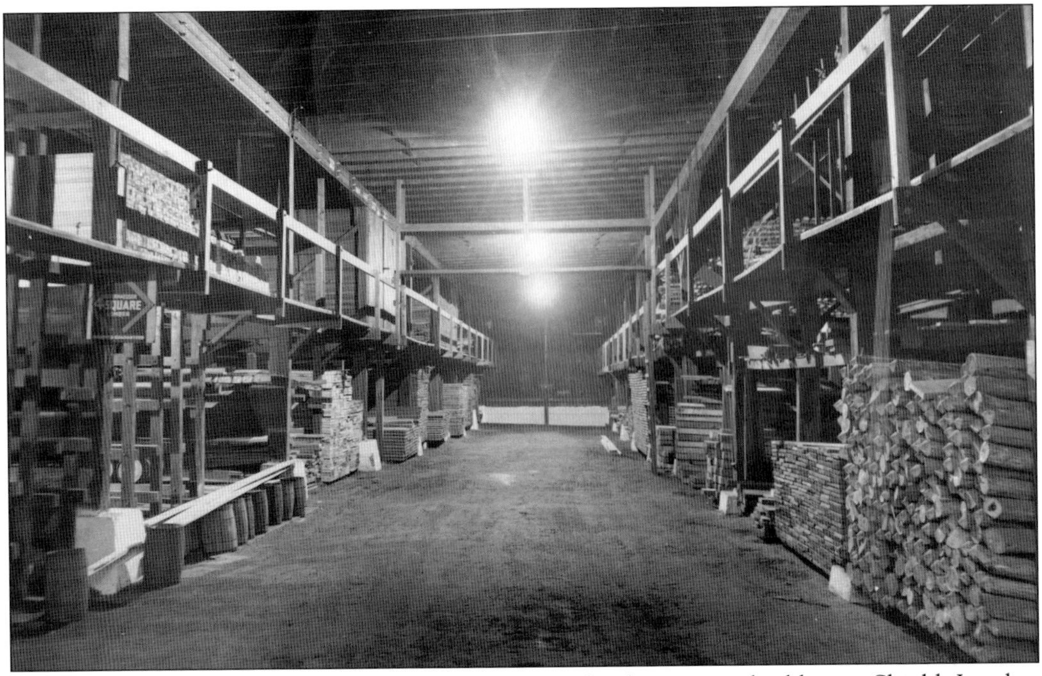

The photograph above shows the interior of the main lumber-storage building at Shields Lumber & Coal in Greenville in 1947. Finished lumber and fencing material can be seen lining both sides of the building, which was just north of the office and store, with kegs of nails lining the floor at left. The photograph below shows another view of the main lumber shed with views of the stock of white pine and cypress boards. Shields was popular for both contractors and local estate managers because of their wide array of products in stock that were ready for pickup. They stocked so much lumber that when they began to reorganize the lumber stockpile in 1930, the job took nine years. (Both, courtesy of John Shields.)

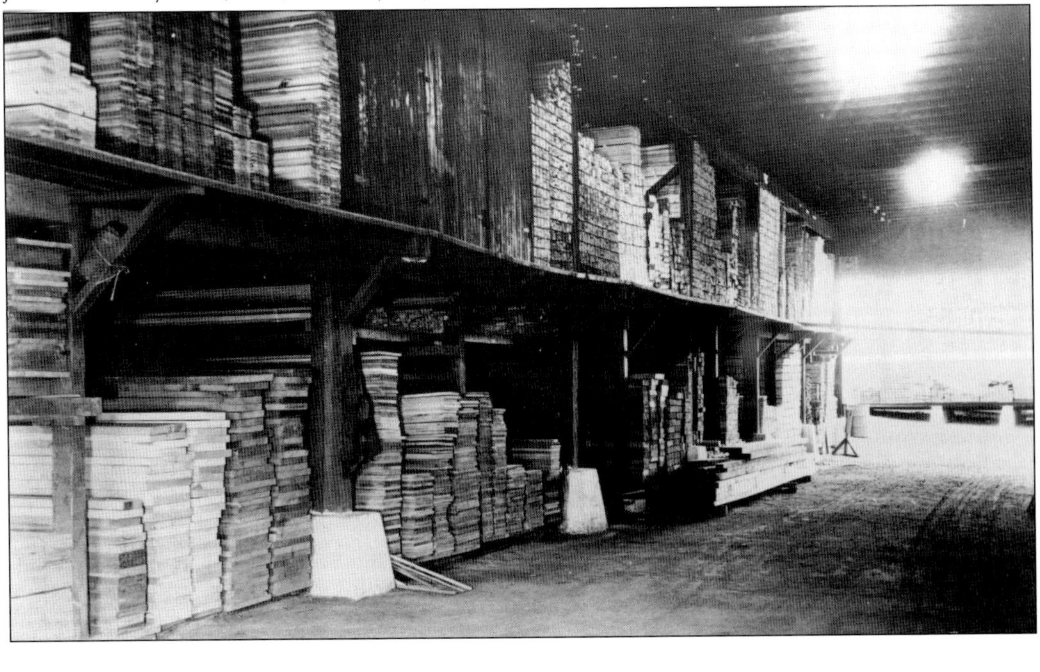

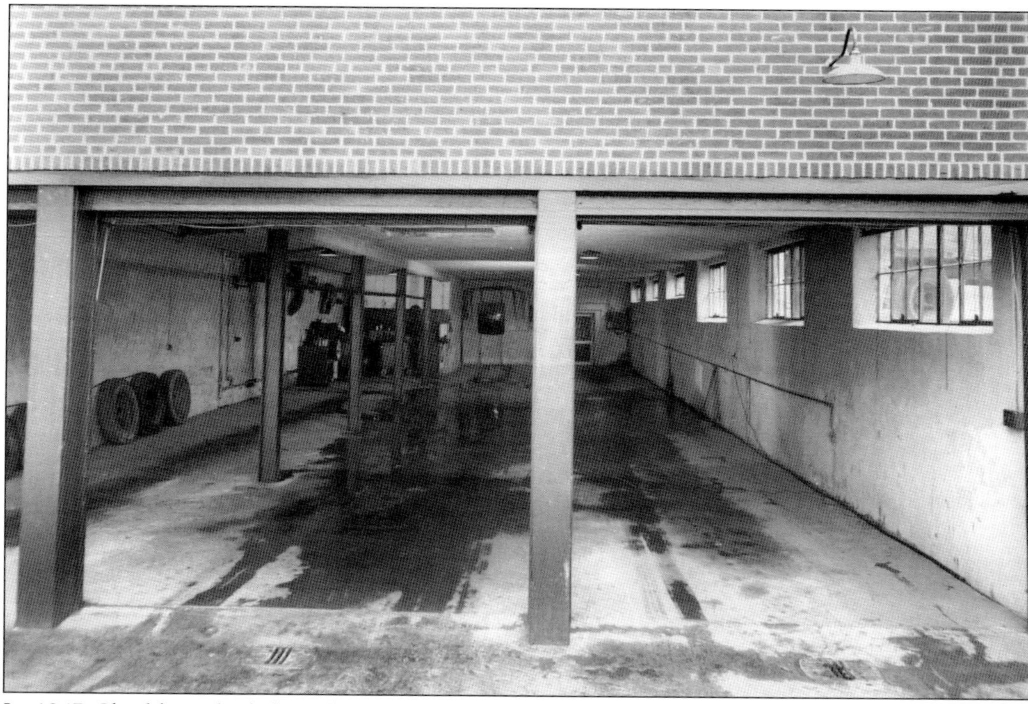

In 1947, Shields parked their fleet of cars and trucks under the main warehouse along Kennett Pike. Not only did the firm have a fleet of delivery trucks, but it also had a few business cars for the use of salesmen attending to customers on the job. The main warehouse was built to the highest standards for building codes and safety and included fire doors that separated the basement garage and the storage areas. This served a dual purpose that allowed for loading materials during inclement weather without damage. (Both, courtesy of John Shields.)

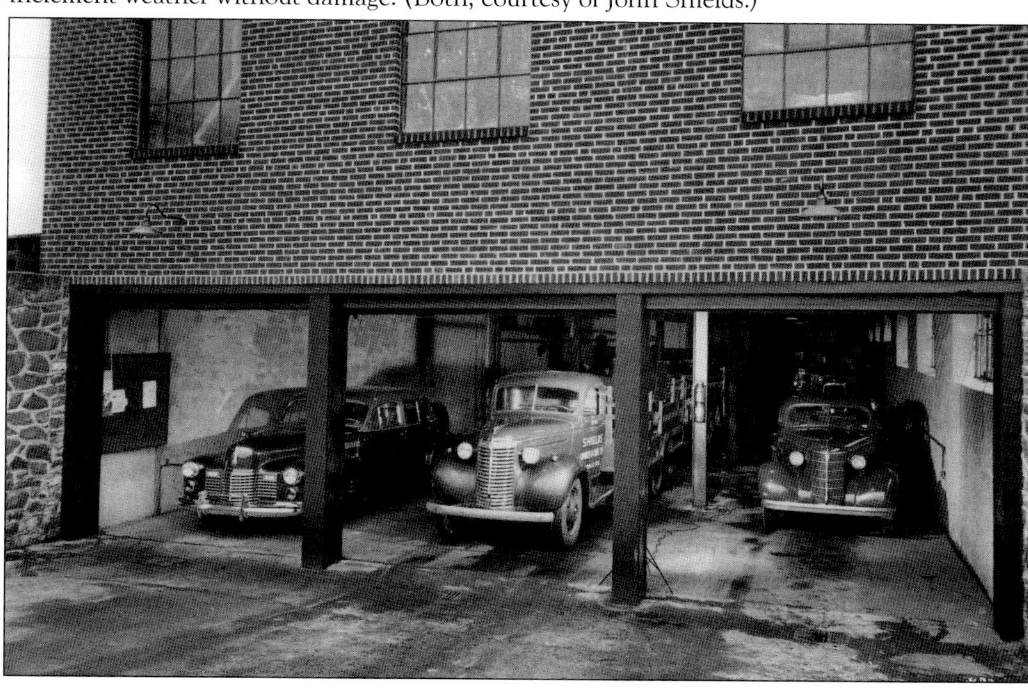

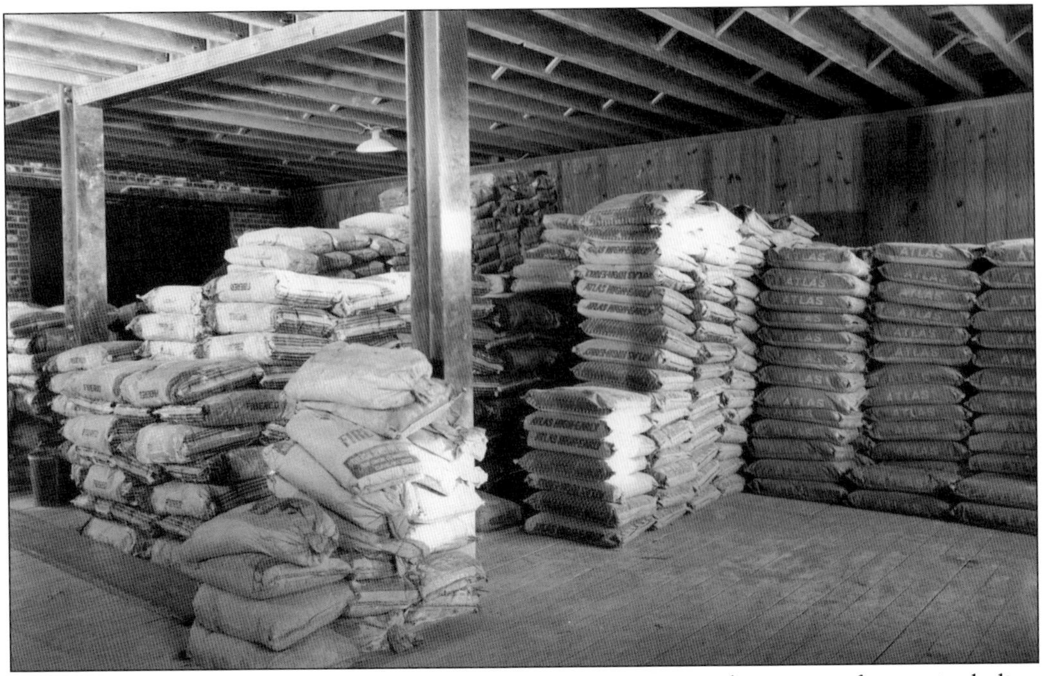

The first floor of the warehouse building, shown here in 1947, stored a variety of items, including cement, plaster, lime, and landscaping products, such as grass seed. Built in 1936, it was constructed for the storage of masons' material, millwork, insulation, and seasoned lumber. The building itself featured many of the materials Shields sold so customers could see them in their actual use. When the new store and office structure was built in 1940, Shields earned the reputation of having one of the best layout and management styles in the business nationwide. In fact, they were awarded second place in a national contest run by the *American Lumberman*. (Both, courtesy of John Shields.)

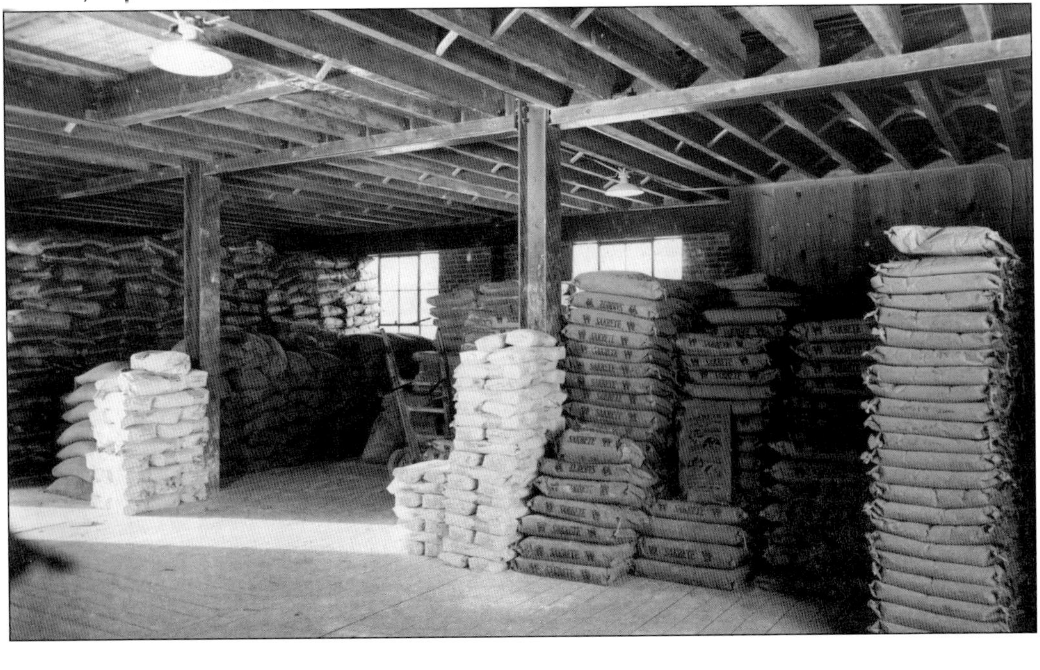

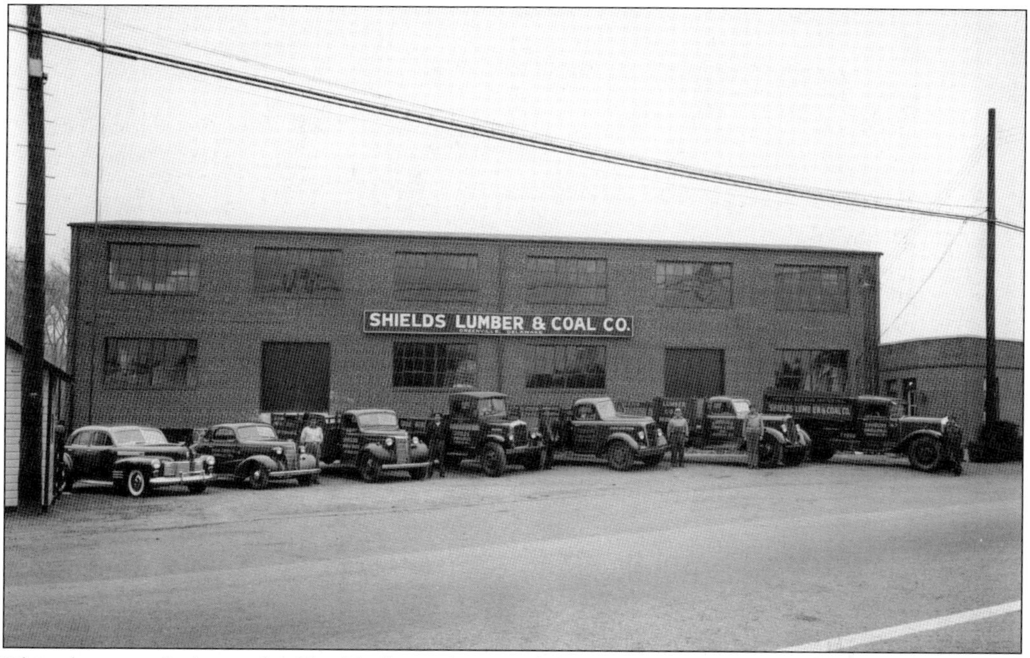

This photograph shows the Shields fleet of automobiles and delivery trucks in front of the second warehouse, which was built by Daniel F. Shields Jr. along Kennett Pike in 1936. Kennett Pike is in the foreground. This building was located just across from the intersection of Kennett Pike and Hillside Road. (Courtesy of John Shields.)

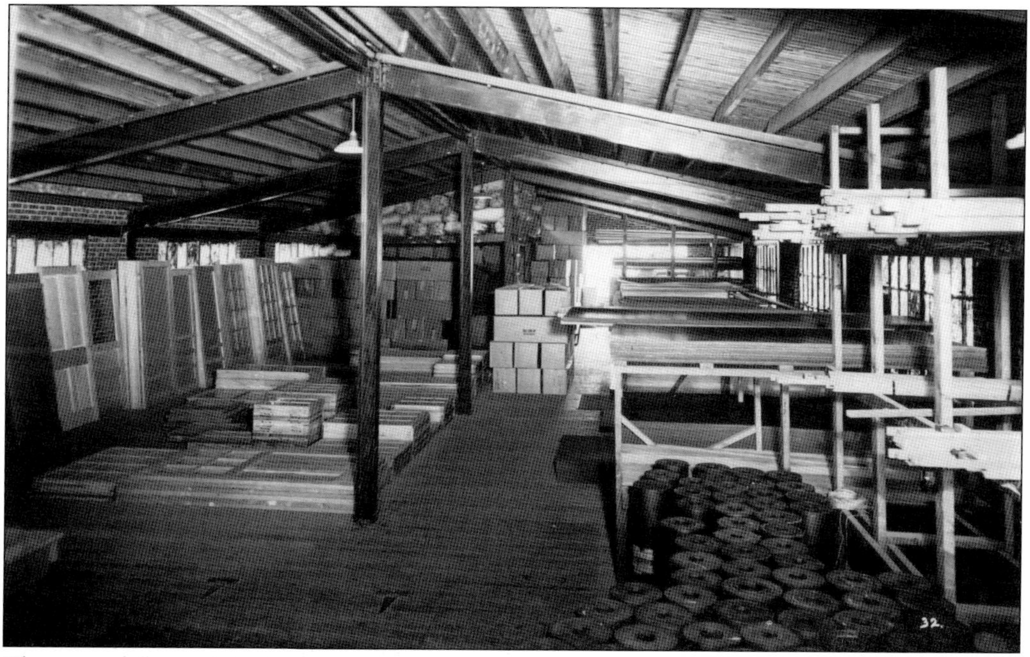

The second floor of the Shields warehouse contained a wide variety of construction materials. Here, everything from millwork and roofing materials to windows and doors were kept stocked for the convenience of customers in Greenville and the surrounding estates. Many local estates ran charge accounts with the retailer. (Courtesy of John Shields.)

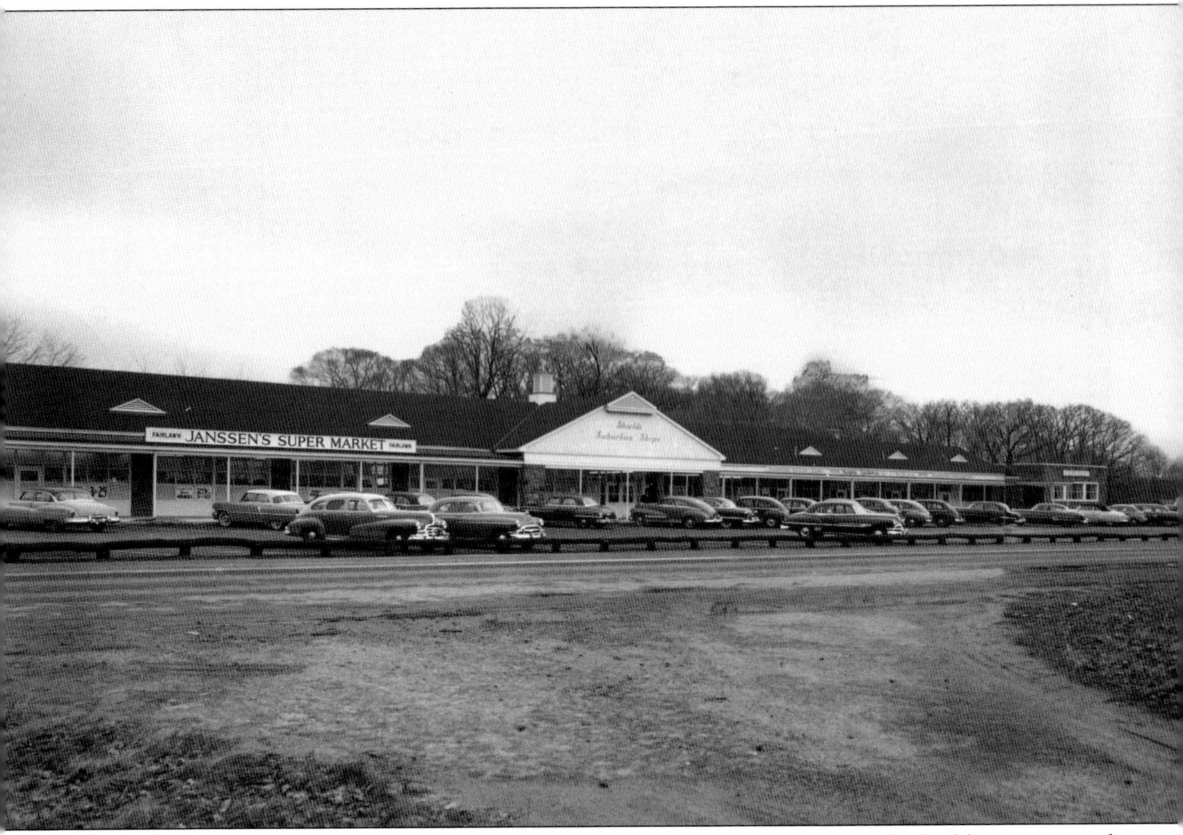

With the great success of the Shields retail store along Kennett Pike, Daniel Shields constructed the Shields Suburban Shoppes in 1950. The strip center was the first modern shopping center in the Greenville area. The view above shows the Shoppes in approximately 1966, looking south toward Wilmington. Many of the original tenants from 1950 remained in the shopping center well into the late 1960s with the exception of the Greenville branch of the federal post office. When Daniel Shields developed the complex, he brought his three sons into the new venture called the Shields Development Company. By the mid-1960s, Daniel F. Shields III was its vice president, William Shields, its treasurer, and John J. Shields Sr., its secretary and general manager. The Shields family continued to own the parcel until the late 1980s, when the hardware store and land were sold to a local real estate management firm. (Courtesy of Hagley Museum and Library.)

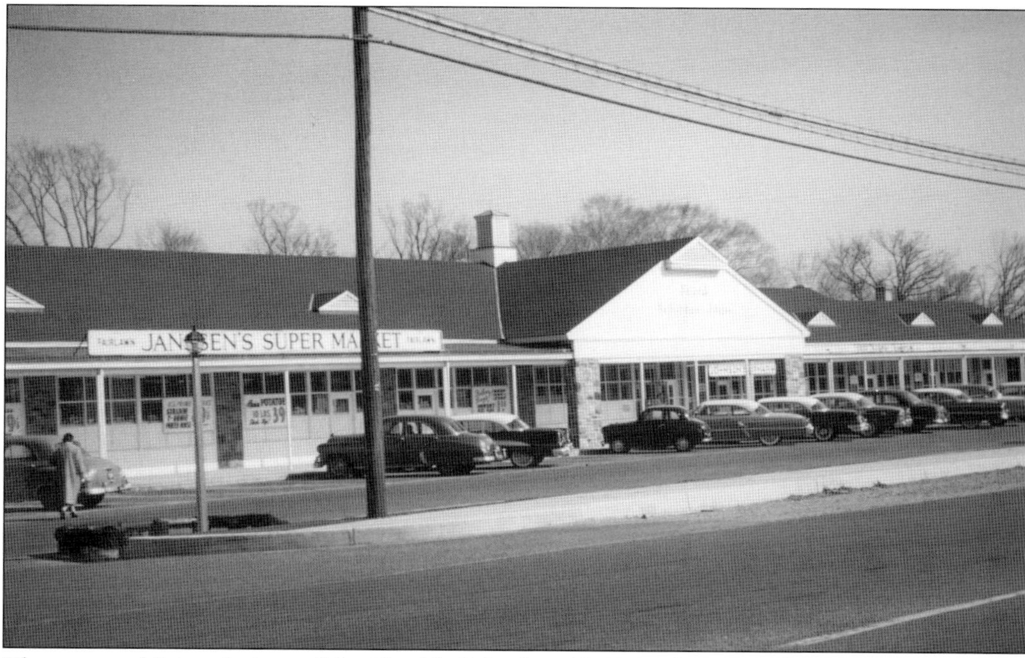

These two additional views of the Shields Suburban Shoppes show storefronts and Kennett Pike in the foreground. The image above shows Janssen's Market (now located in Greenville Crossing and owned by the Janssen family) in the mid-1950s. The image below shows an aerial view of the Shoppes in the early 1960s before the addition of what would become the Shields retail store by 1966. Another popular establishment was Johnson's Pharmacy and lunch counter, which was located to the right of Janssen's. Wilmington Country Store, one of the original occupants of the center, is still located in the Shoppes, currently known as Greenville Crossing. (Both, courtesy of Hagley Museum and Library.)

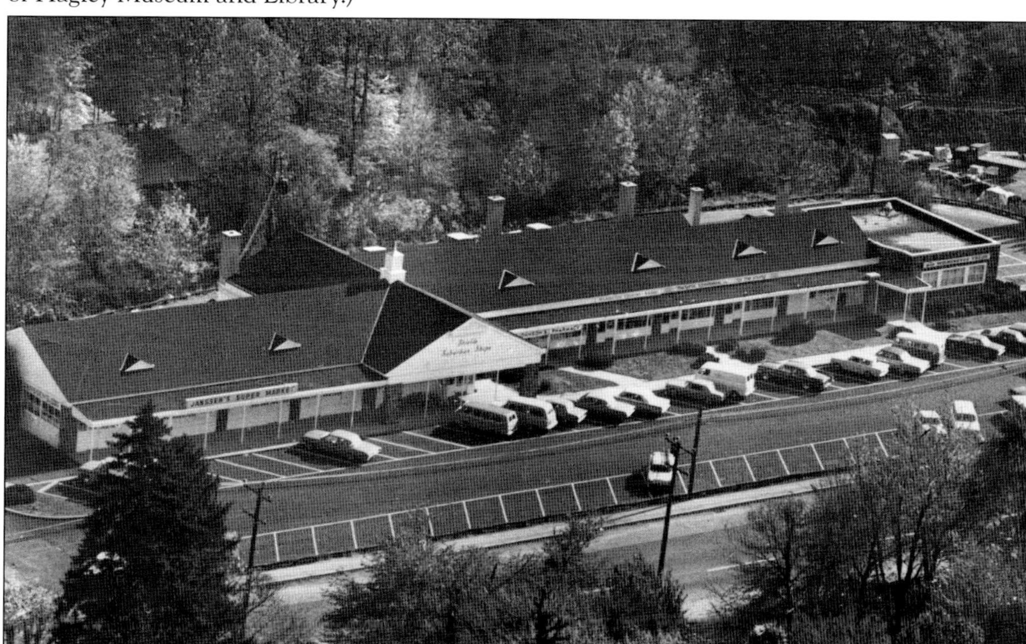

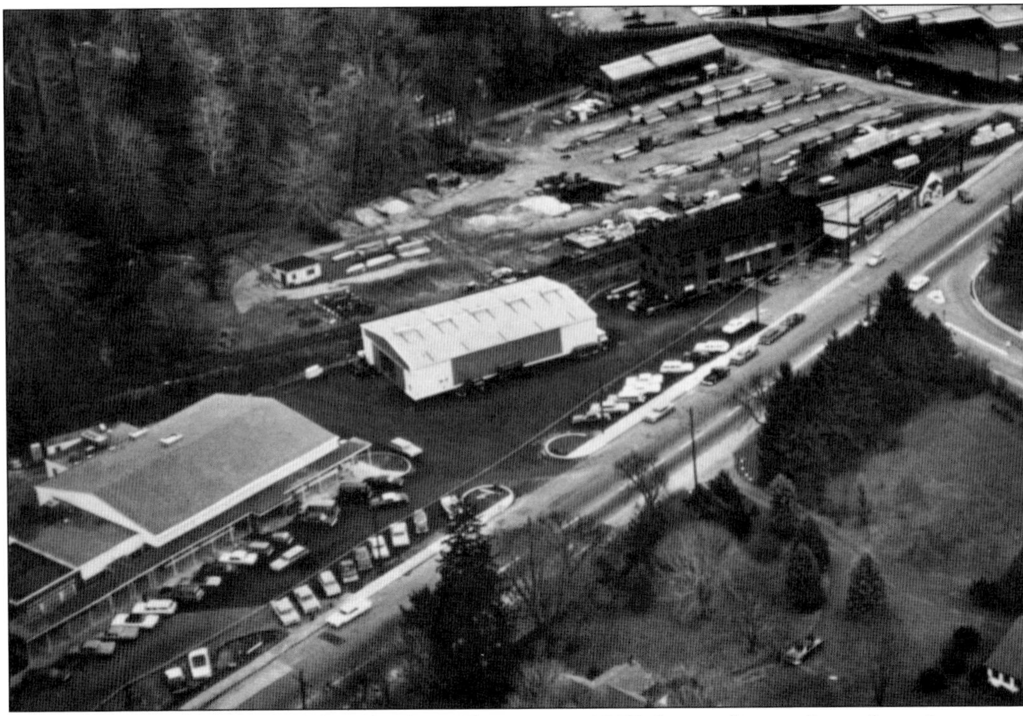

An aerial view of the Shields Suburban Shoppes in 1967 shows the modern shopping center as well as the newly constructed lumber-storage building. The first post office in Greenville is visible at far right; by this time, it was located in the Greenville Center. The background of the image shows the storage yard of the firm, with Hillside Road intersecting Kennett Pike at right. (Courtesy of Hagley Museum and Library.)

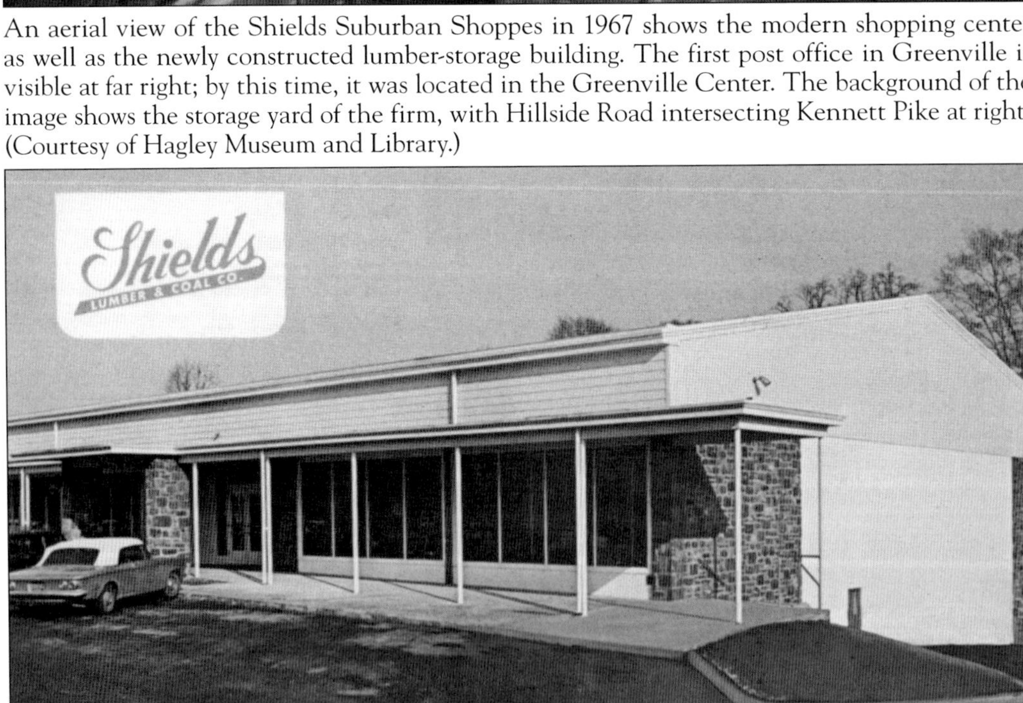

In 1966, when Shields decided to replace the old lumber storage-building with a newer, more modern facility, they also added a new retail store to the end of the Shields Suburban Shoppes, which many in the area remember as a Greenville landmark. (Courtesy of Hagley Museum and Library.)

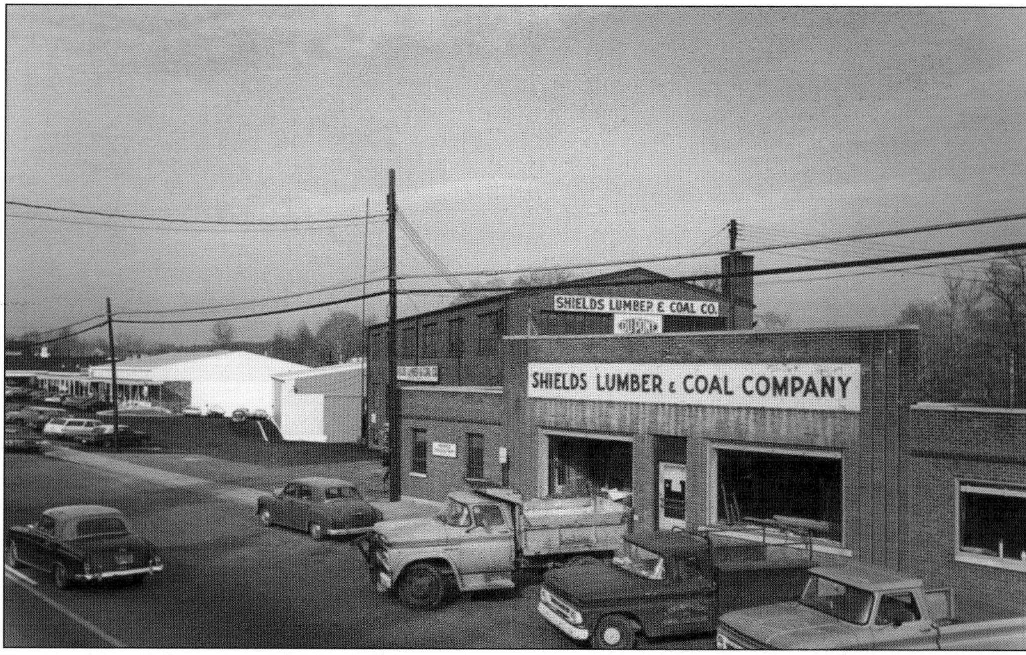

Two views of the new Shields retail store around 1967 show the transformation from the old brick building to the new space within the Shields Suburban Shoppes. The new lumber-storage building can also be seen above; it was constructed around the same time as the store moved to its new location, where it remained into the late 1980s. Following its closure, John Shields Jr. (grandson of the founder) reopened the family business in Hockessin, Delaware, along Lancaster Pike, where it was in operation into the late 1990s. The last surviving son of Daniel F. Shields Jr. passed away in early 2013. (Both, courtesy of John Shields).

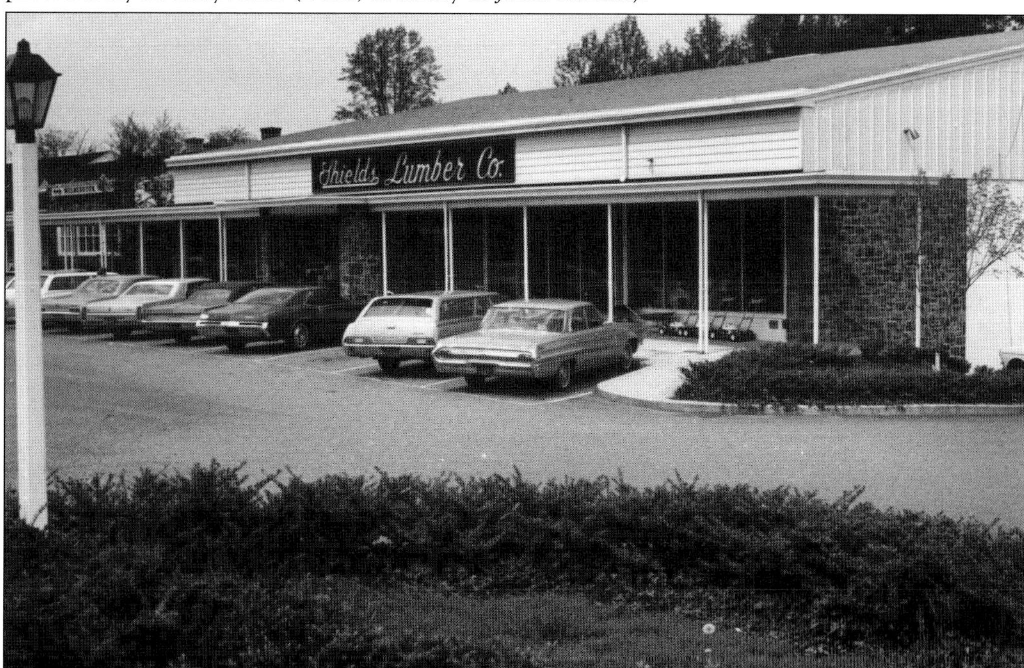

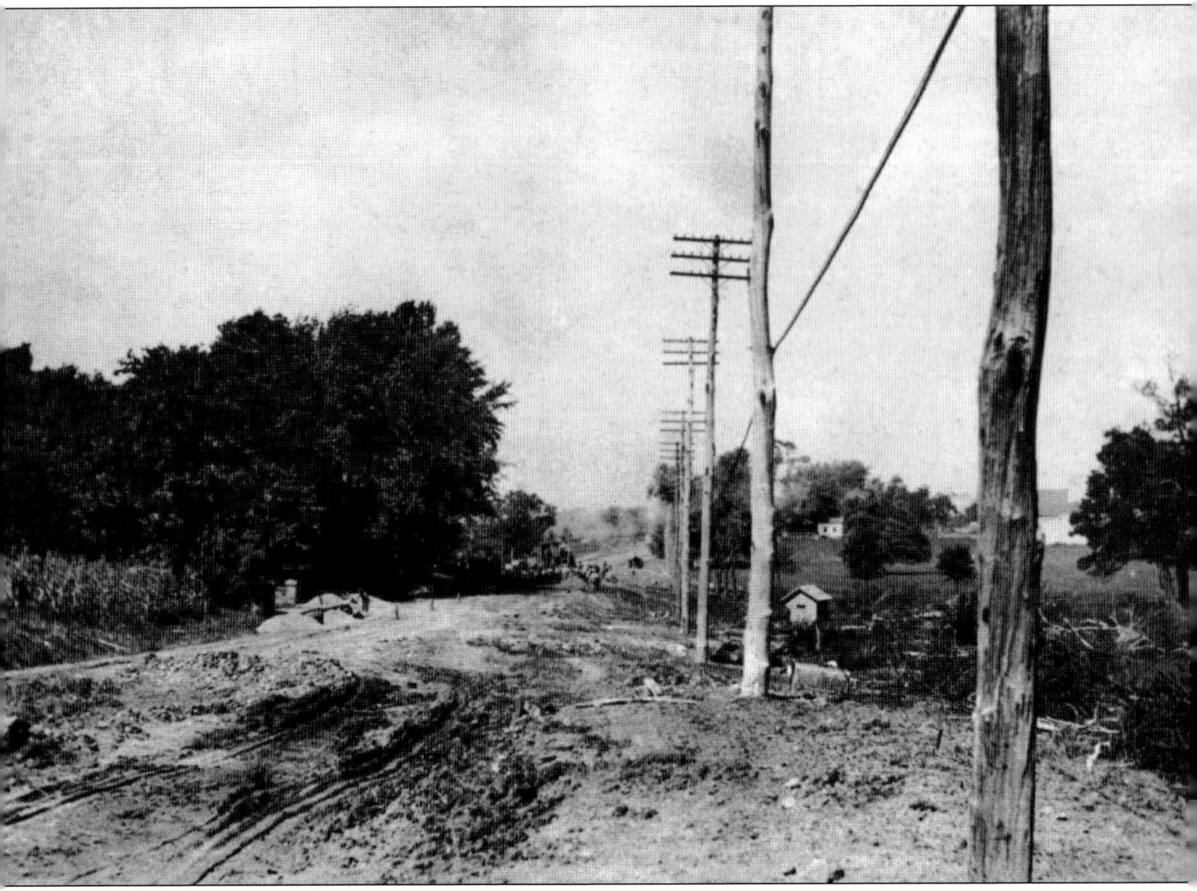

This photograph, taken in 1919, shows Kennett Pike in Greenville during the time of the widening project that was undertaken by Pierre S. du Pont. The view is looking north from near what would later become the Shields property and is now Greenville Crossing, home to the Wilmington Country Store, Walgreens, and BBC Tavern & Grill. At left, one can see the driveway to the Green property, which was later the home of the Strange family. Progress reports provided to Pierre S. du Pont on a weekly basis by the DuPont Company and its engineers noted that the work on this particular stretch of Kennett Pike took place in October 1919. The excavation work in Greenville continued through November and was complete by December. Paving this section was done the following year after the winter weather allowed for work to resume. (Courtesy of John Shields.)

Three

Winterthur Vicinity

After leaving the village of Greenville and heading north along Kennett Pike, travelers cross the intersection of Campbell/Kirk Roads and come to an area in the vicinity of Winterthur. Although the area along the pike is not a part of Wilmington proper, postal addresses are not included as a part of Greenville or nearby Centreville.

The origins of the Winterthur estate date to 1810 when E.I. du Pont, the founder of E.I. du Pont de Nemours & Company, purchased the property in four separate parcels. After du Pont's death in 1834, the property was left to his children. His daughter Evelina bought out the shares owned by her siblings, and she and her husband, Jacques Antoine Bidermann, began construction of a new home, which was completed in 1839. Following the deaths of the Bidermanns in the 1860s, Henry du Pont (Evelina's brother) purchased the property for his son Col. Henry A. du Pont in 1867. By the time of Colonel du Pont's death in 1926, the house had undergone a major renovation and featured more than 2,000 acres, complete with formal gardens, a working farm, and even its own post office. His son Henry Francis du Pont transformed the property into the estate and museum of today, featuring one of the finest collections of early American antiques in the world.

The estate encompassed much of the land on both sides of Kennett Pike, north of the intersection with Campbell Road all the way to the intersection with Pyles Ford Road. Comprising the many farms that various du Ponts bought over a span of a decade, many of the dwellings along Kennett Pike were at one point associated with the workings of the estate.

Today, the holdings of the Winterthur property are less than half of what they once were; however, they have left their mark on the landscape. For example, the Bidermann Golf Club property (once Henry Francis du Pont's private course) and the grounds of Wilmington Country Club were all once a part of its vast holdings along Kennett Pike.

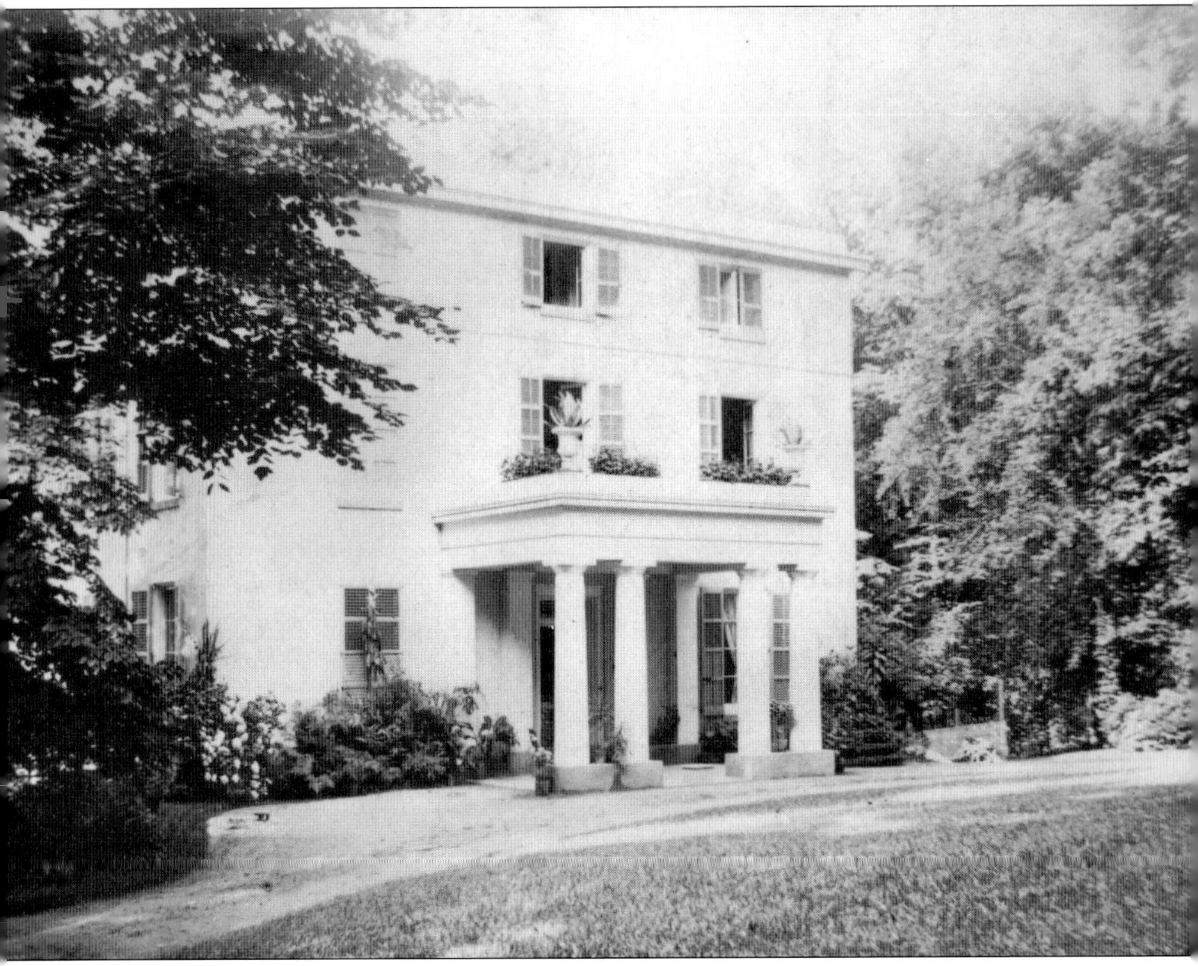

Pictured is Winterthur, shown toward the end of the 19th century during the period of occupancy of Col. Henry A. du Pont. Colonel du Pont had been given the property by his father, Gen. Henry du Pont, and undertook two renovations of the mansion in 1884 and 1902 (the latter included the redesign of the gardens). Col. Henry A. du Pont led an interesting and full life. Born in 1839, he graduated from West Point and saw active combat during the Civil War. After the war, he was president of the Wilmington & Kennett Turnpike Company and served on the boards of directors for DuPont and the Reading Company. In 1906, he was elected to the US Senate, where he served until 1916. (Courtesy of Hagley Museum and Library.)

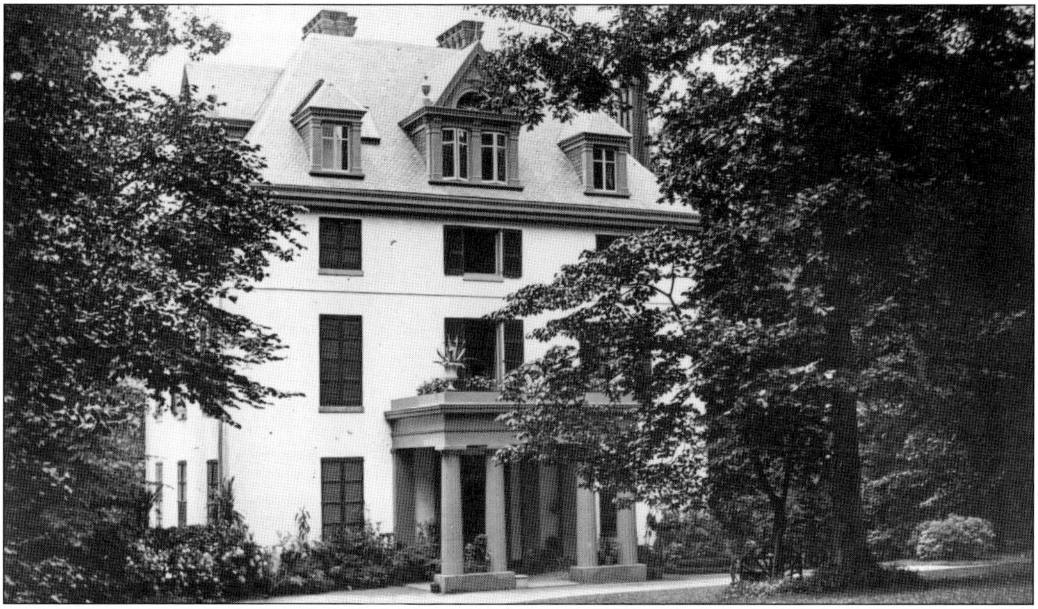

These two photographs show Winterthur during the occupancy of Col. Henry A. du Pont. Seated on the lawn in the image below around 1884 are Colonel du Pont's children Louise (later Crowninshield) and Henry Francis. Henry Francis inherited the mansion and surrounding property following the death of his father in 1926. He undertook huge renovations during the period between 1929 and 1931. Not only did he enlarge the main house to include a squash court and bowling alley, he also greatly expanded the estate gardens, many of which can be visited today by the public. (Both, courtesy of Hagley Museum and Library.)

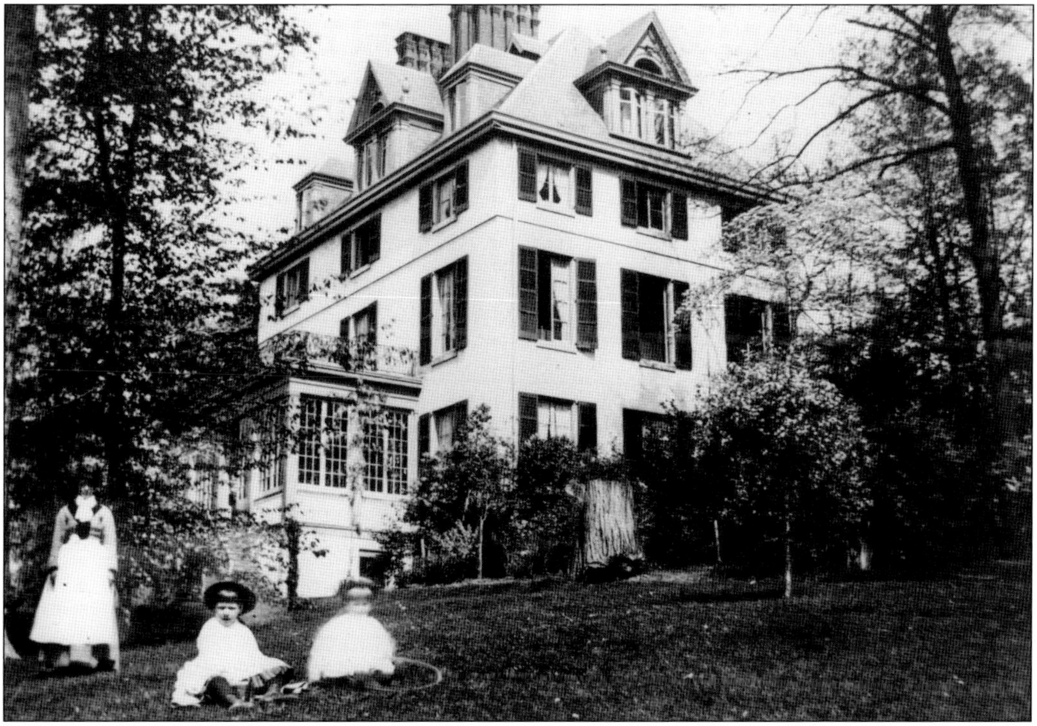

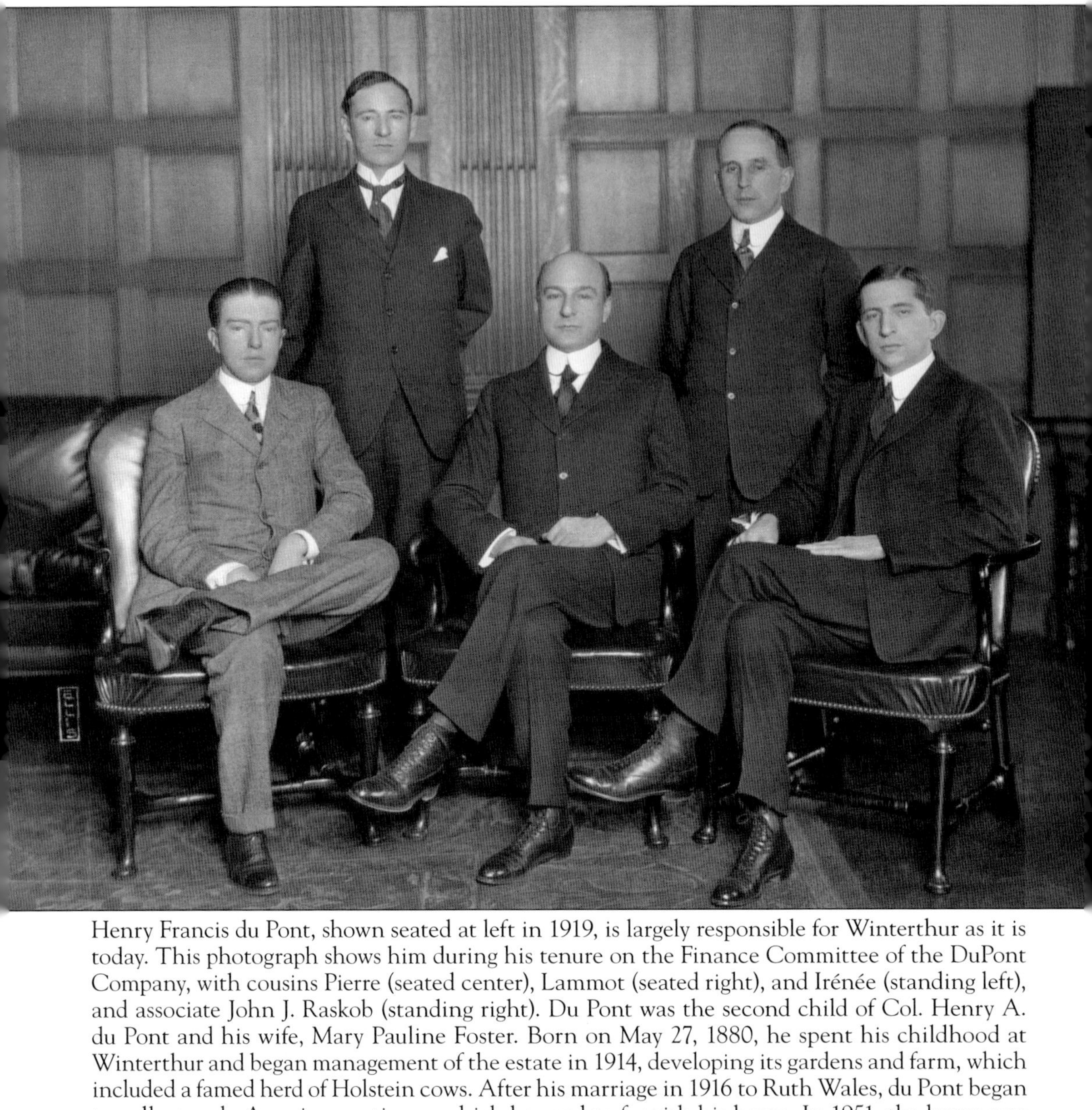

Henry Francis du Pont, shown seated at left in 1919, is largely responsible for Winterthur as it is today. This photograph shows him during his tenure on the Finance Committee of the DuPont Company, with cousins Pierre (seated center), Lammot (seated right), and Irénée (standing left), and associate John J. Raskob (standing right). Du Pont was the second child of Col. Henry A. du Pont and his wife, Mary Pauline Foster. Born on May 27, 1880, he spent his childhood at Winterthur and began management of the estate in 1914, developing its gardens and farm, which included a famed herd of Holstein cows. After his marriage in 1916 to Ruth Wales, du Pont began to collect early American antiques, which he used to furnish his home. In 1951, the house was opened to the public, and the family moved to smaller home just steps away from the mansion. Du Pont later was the head of the Fine Arts Committee, which worked with Jacqueline Kennedy to restore the White House. (Courtesy of Hagley Museum and Library.)

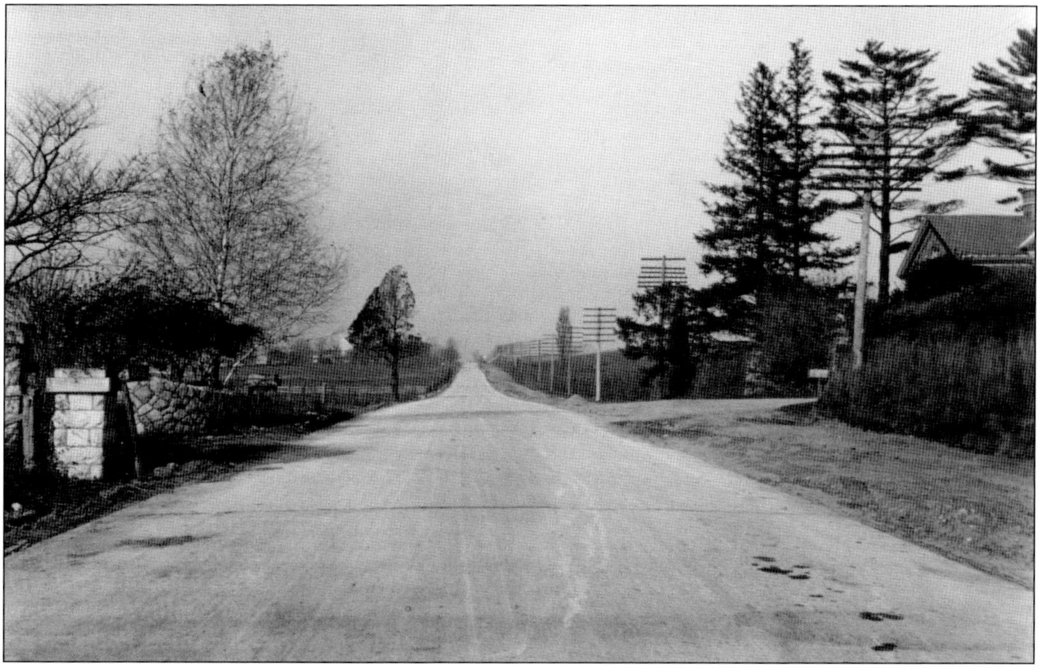

Two views looking north on Kennett Pike around 1917 show the width of the road before the widening project undertaken by Pierre S. du Pont in 1920. The building shown at the right in both images is the gatehouse and original entrance to Winterthur. Today, the Methodist Country House retirement village is located to the left. At right, just south of the gatehouse, is the entranceway to the Wilmington Country Club. In recent years, the gatehouse has been used as a ticketing office for the estate's popular steeplechase event, Point to Point. (Both, courtesy of Hagley Museum and Library.)

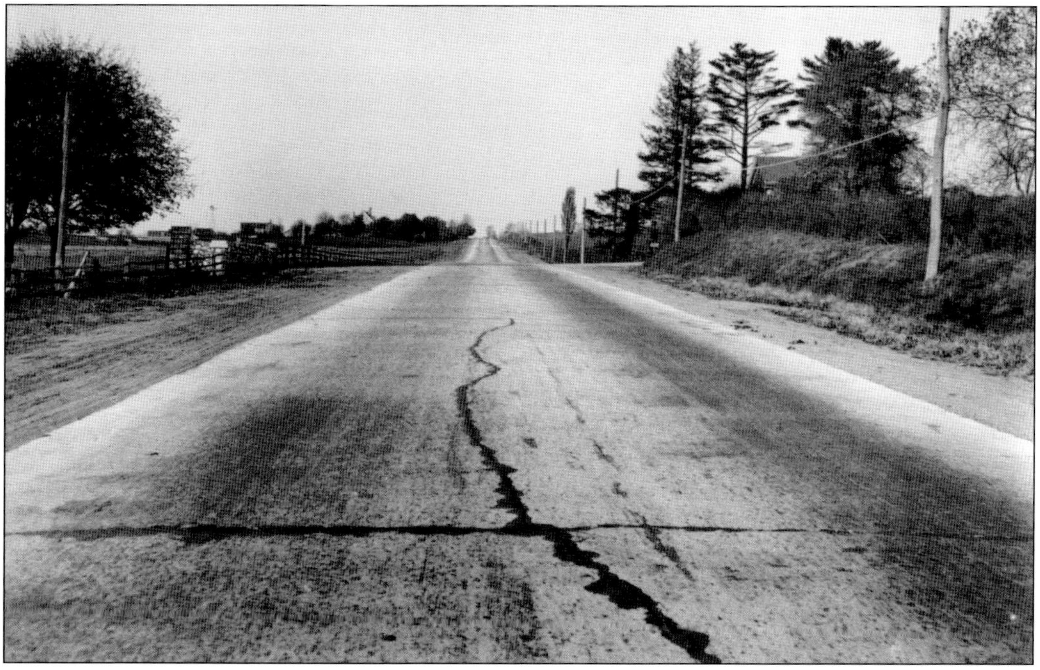

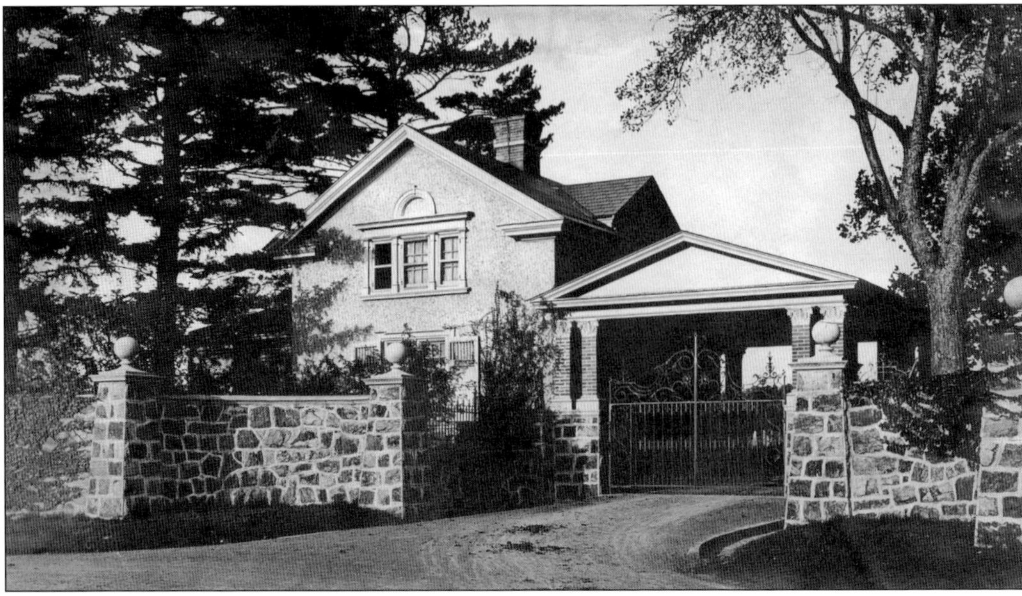

This photograph shows the gatehouse at Winterthur. Although undated, the photograph was likely taken in the early part of the 20th century. This gate served as the main entrance to the estate until the 1960s, when a new entrance was constructed farther north on Kennett Pike. (Courtesy of the Winterthur Library: Winterthur Archives.)

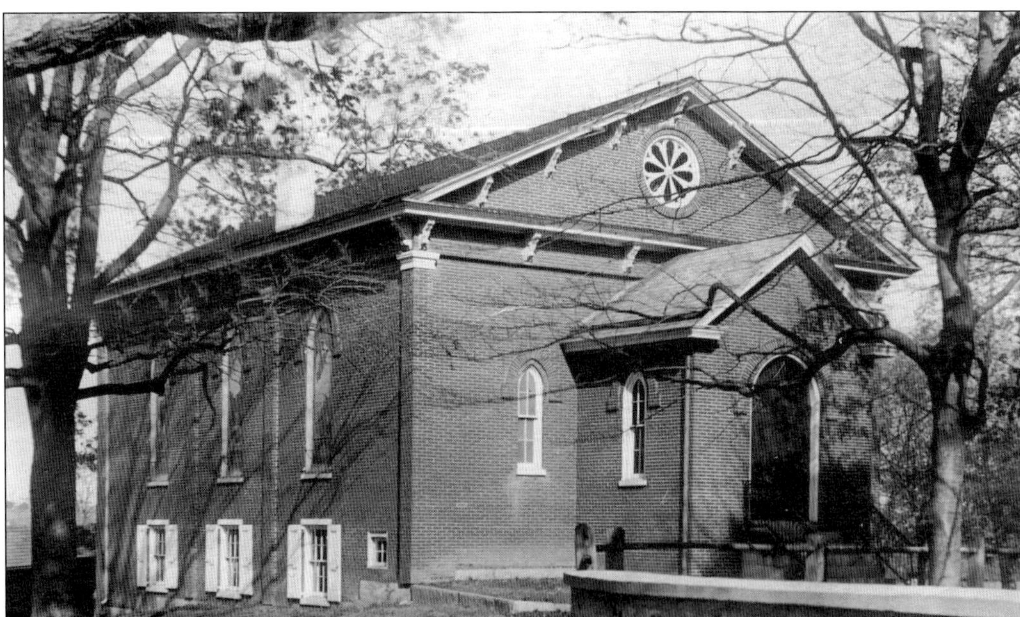

Located across from the present-day entrance to Winterthur is Lower Brandywine Presbyterian Church. Tracing its roots to 1720, the current building dates to the mid-19th century; it replaced an old log structure that was on the property. Many records from the period refer to Lower Brandywine as "the old brick church." (Courtesy of Hagley Museum and Library.)

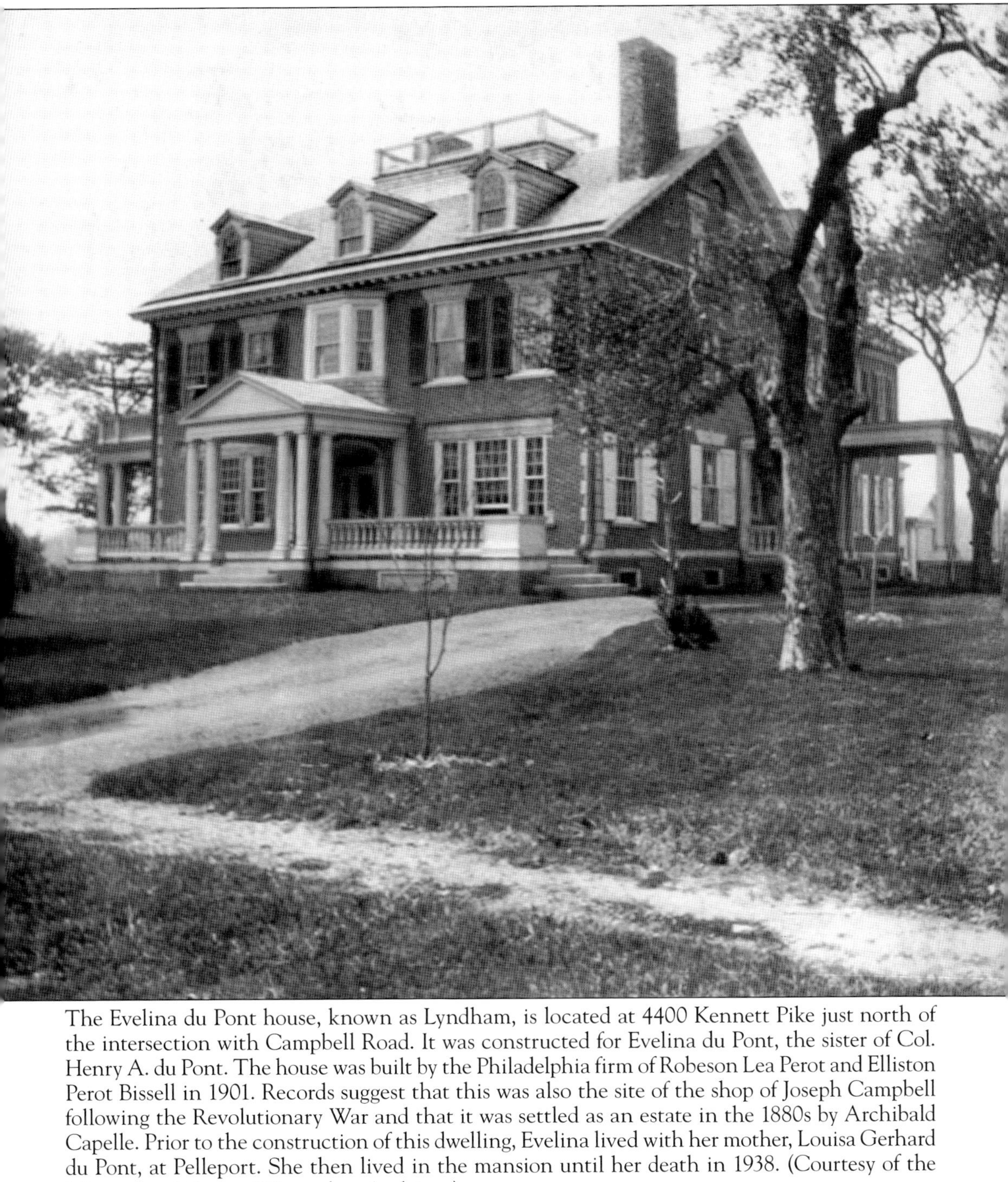

The Evelina du Pont house, known as Lyndham, is located at 4400 Kennett Pike just north of the intersection with Campbell Road. It was constructed for Evelina du Pont, the sister of Col. Henry A. du Pont. The house was built by the Philadelphia firm of Robeson Lea Perot and Elliston Perot Bissell in 1901. Records suggest that this was also the site of the shop of Joseph Campbell following the Revolutionary War and that it was settled as an estate in the 1880s by Archibald Capelle. Prior to the construction of this dwelling, Evelina lived with her mother, Louisa Gerhard du Pont, at Pelleport. She then lived in the mansion until her death in 1938. (Courtesy of the Winterthur Library: Winterthur Archives.)

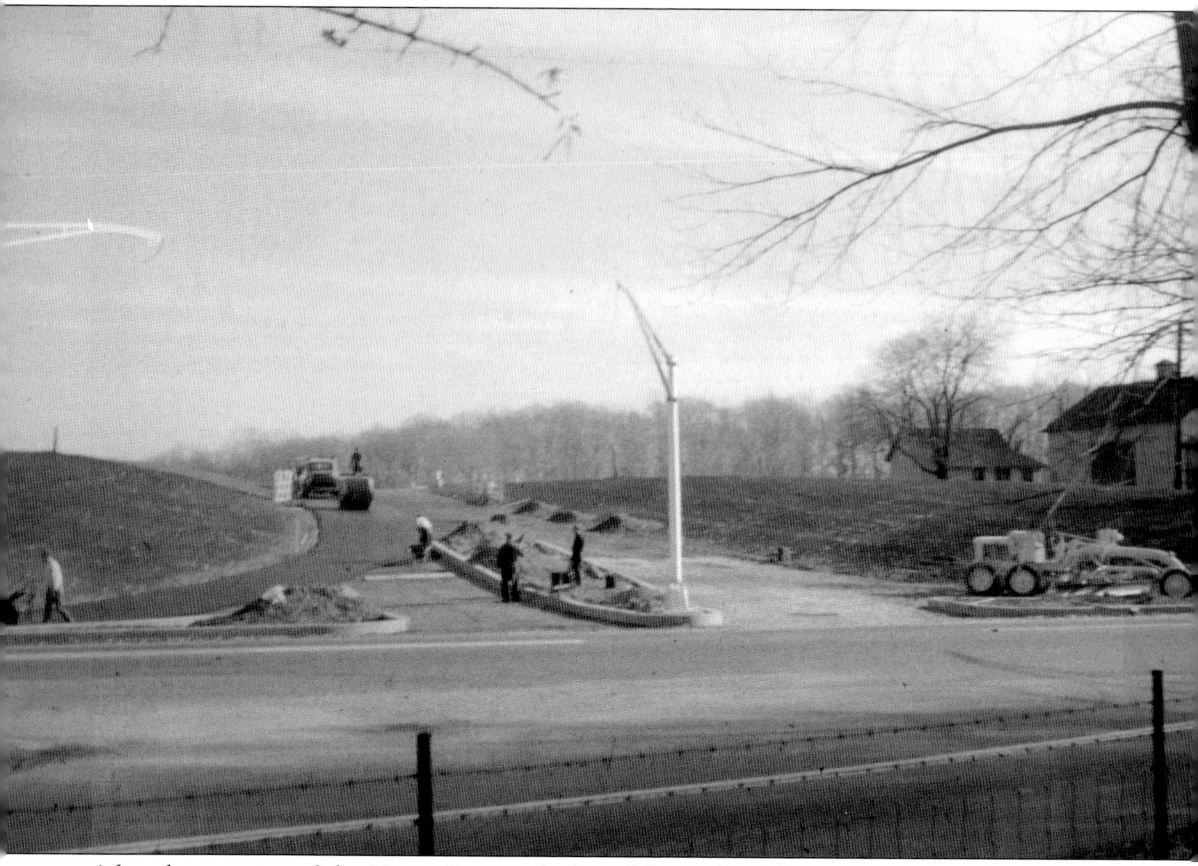

After the opening of the Henry Francis du Pont Winterthur Museum in 1951 and because of increased visitation throughout the 1950s, a new entrance to the estate was added north of the gatehouse and across from the intersection of Kennett Pike and Old Kennett Road. Work was begun in 1959 on the new entrance and was completed in 1960. It continues to serve as the main entrance to the property for visitors. The museum and its extensive collection of early American antiques are renowned worldwide. (Courtesy of the Winterthur Library: Winterthur Archives.)

This photograph shows the Raley Missile House, which was constructed in 1961 by Winterthur carpenters at the new entrance to the museum. For nearly 45 years, the octagonal structure stood along Kennett Pike, near its intersection with Old Kennett Road, greeting visitors and travelers. (Courtesy of the Winterthur Library: Winterthur Archives.)

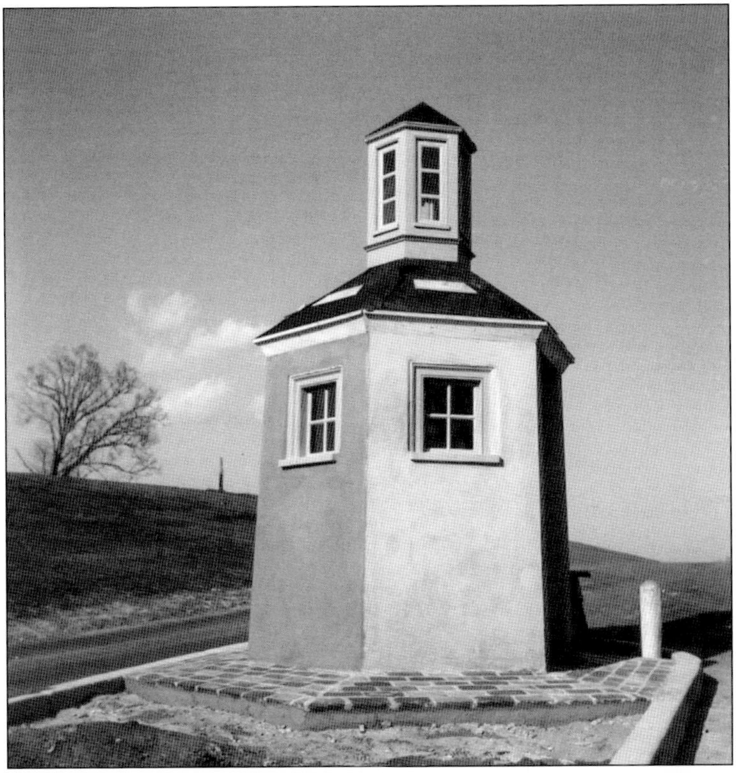

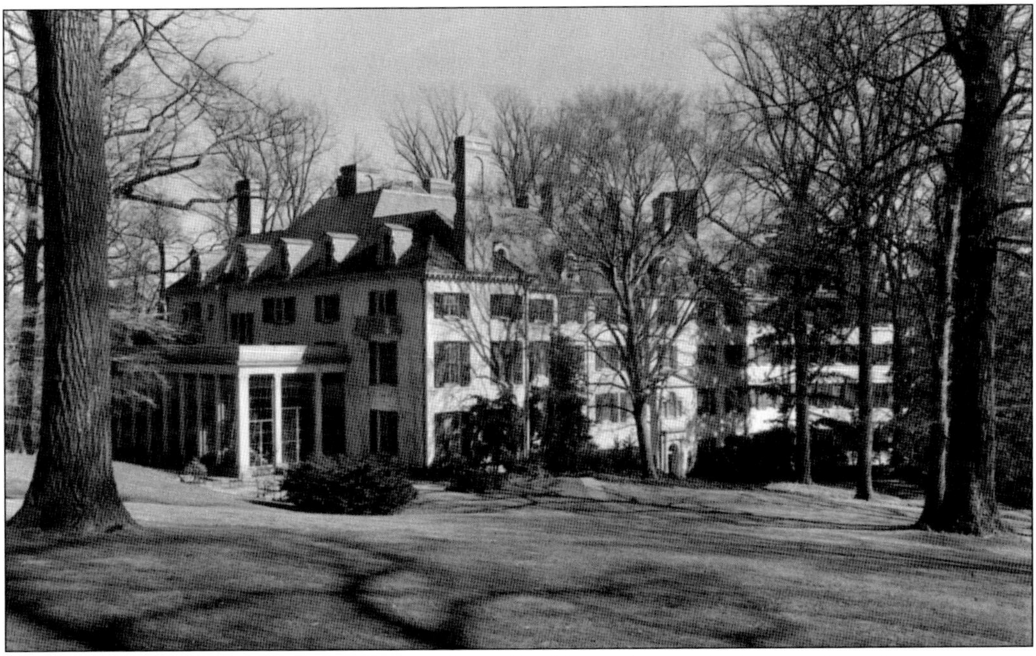

By the late 1940s, Winterthur had been enlarged by Henry Francis du Pont and was being prepared to become a museum. This photograph shows the northeast corner of the house, likely in the 1930s. It remained unchanged until the addition of the south wing in the mid-1950s. (Courtesy of the Winterthur Library: Winterthur Archives.)

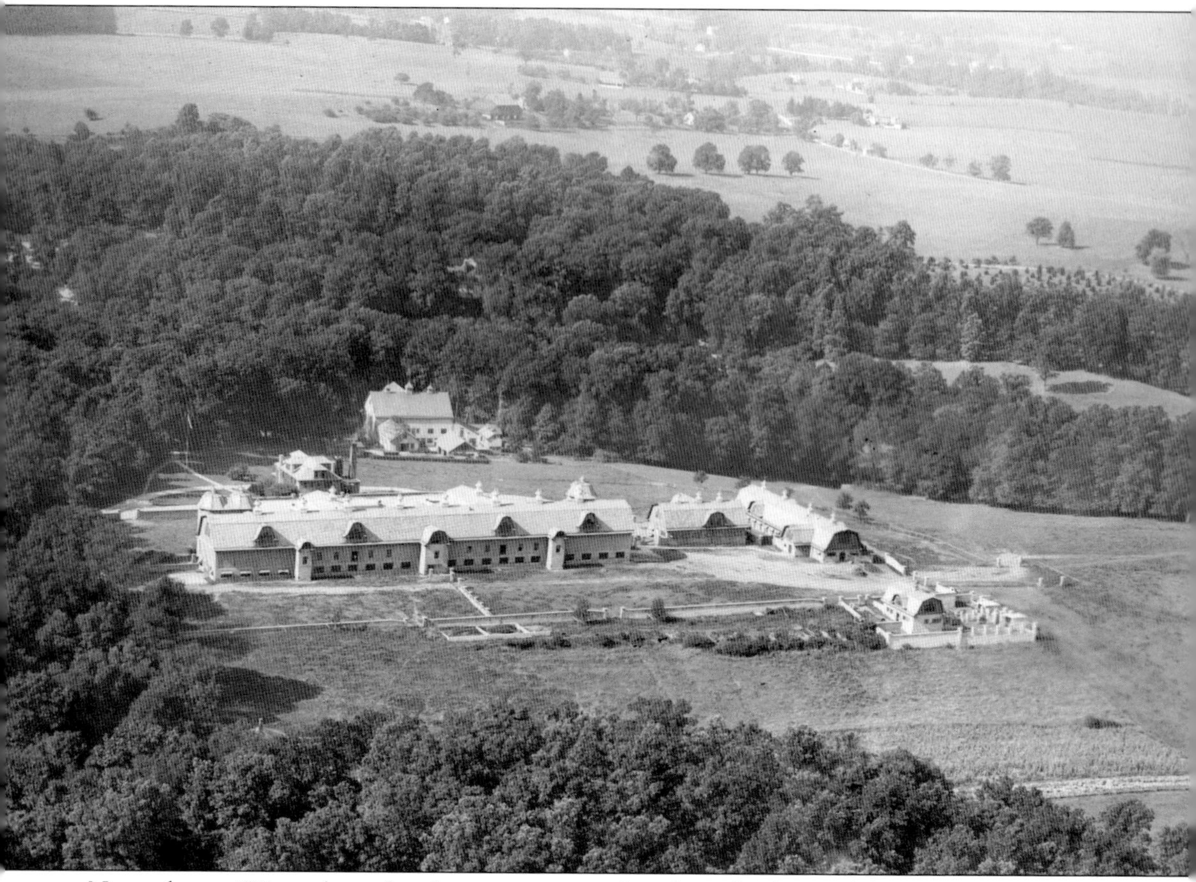

Not only was Winterthur a lavish country estate for a branch of one of the country's wealthiest families, it was also a working farm. For more than a century, it was a self-sustaining estate. H.F. du Pont raised a world-famous herd of Holstein cows using what were at the time the most-advanced means of herd development. This photograph shows the barn complex located near the mansion (built in the early 20th century). The main barn (center) burned to the ground in 1930. Winterthur has a long history with farming and livestock. When E.I. du Pont purchased the land that is now Winterthur in 1810, he began to raise merino sheep on the property. One of the land's inhabitants was a merino ram named Don Pedro who had come from France in 1801. By 1812, Don Pedro's herd had grown to over 800 sheep. When Don Pedro died in 1811, E.I. du Pont received a condolence letter from Thomas Jefferson. (Courtesy of the Winterthur Library: Winterthur Archives.)

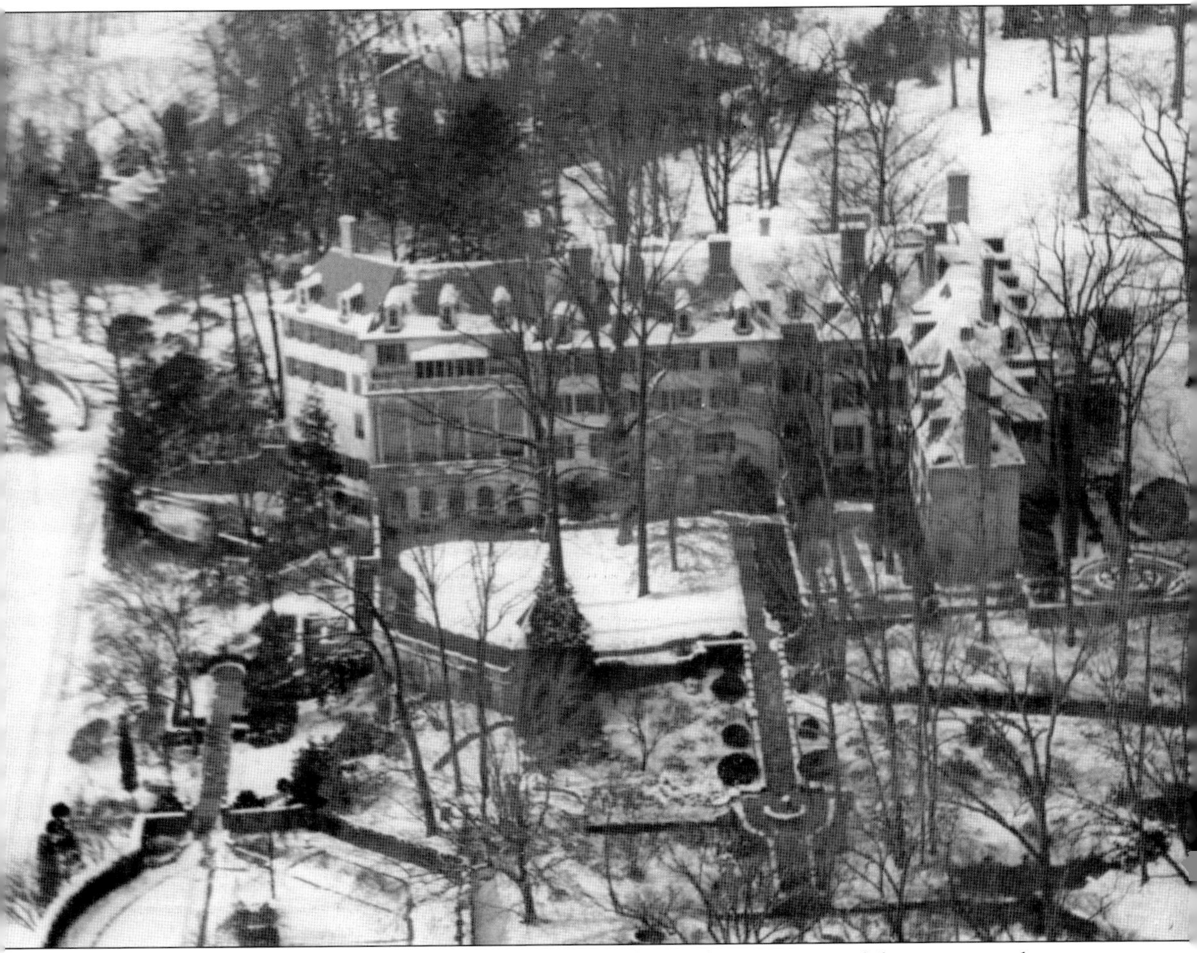

This aerial view of Winterthur, taken around 1930, shows the enormity of the mansion that was built by three generations of du Ponts over the span of more than a century. By 1941, this mansion stood over nine stories and contained more than 140 rooms. Also visible in this snow-covered view are some of the formal gardens. These were designed by Marian Coffin, who also designed the gardens at Gibraltar and St. Amour. The interior of the house featured 11 bedrooms (an entire floor was devoted to guest rooms) as well as multiple parlors and extensive service areas, which included a room to clean shoes, one to clean silver, and many storage areas. The Great Depression hit du Pont hard enough to curtail spending and change his original plans, which had been to open the house as a museum after his death. Instead, the museum opened in 1951 to its first visitors while du Ponts lived in a nearby residence. (Courtesy of the Winterthur Library: Winterthur Archives.)

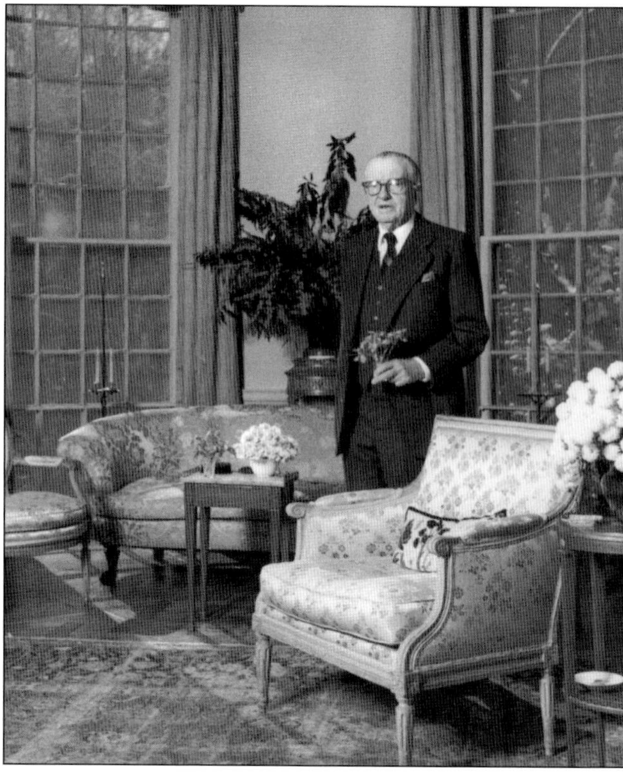

This 1969 photograph shows Henry Francis du Pont in his office at the "Cottage," where he and his family moved after the main house opened as a museum. Just steps away from the mansion, the building was renovated before the family moved in and was decorated with foreign furnishings that could not be used in the museum. (Courtesy of the Winterthur Library: Winterthur Archives.)

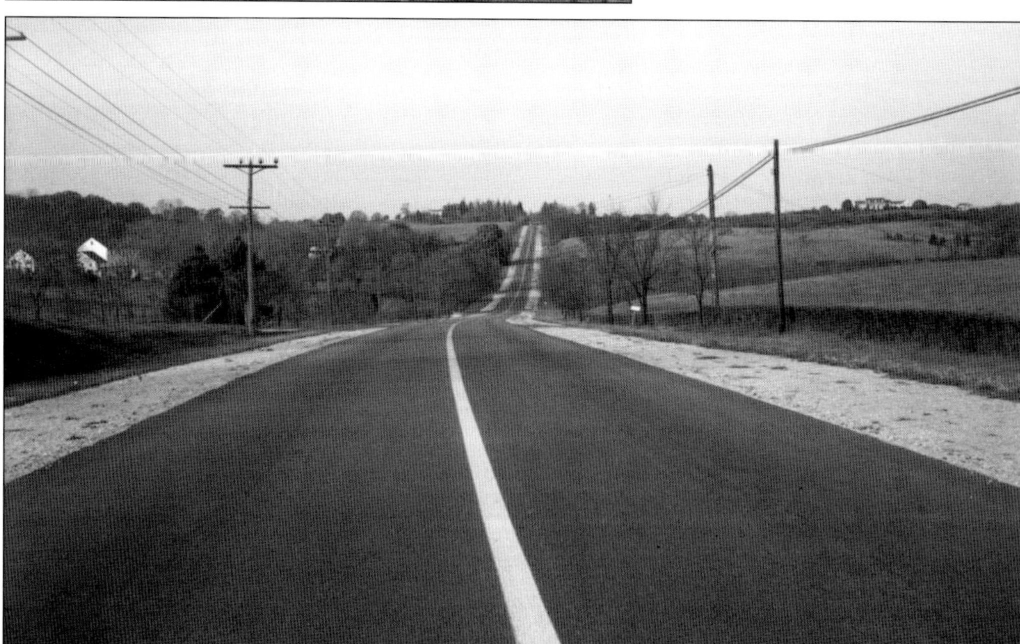

The view looking north from the vicinity of Lower Brandywine Presbyterian Church in 1942 toward the village of Centreville shows the widened Kennett Pike with its unpaved shoulders. At the bottom of the hill is Pyle's Ford Road, a local cut-through to Centre Meeting Road that runs along the present-day border of Winterthur. (Courtesy of the Delaware Public Archive.)

Four

Centreville

Following Route 52 north past Winterthur, travelers soon come to the historic village of Centreville, Delaware. The earliest settlers came to Centreville around 1680; however, the founding of the village dates to 1750. The village center continued to grow, and by the middle of the 19th century, Centreville was home to two inns, a general store, a post office, and even a few pubs. While the DuPont Company was undergoing tremendous expansion in the second half of the 1800s, many du Pont family members began to move from the banks of the Brandywine River north to Centreville and its surrounding countryside. This was the beginning of what is now called "Chateau Country."

When Pierre S. du Pont bought all of the shares of the Wilmington & Kennett Turnpike Company and modernized Kennett Pike, it further encouraged the movement of du Pont family members, employees, and others to the rolling hills of Centreville. In fact, in the period between 1918 and 1920, du Pont family members and their respective business interests owned 48 percent of the road's Delaware frontage, by default halting the suburbanization experienced by many other areas of northern New Castle County. In 1937, Centreville was the site of the wedding of Ethel du Pont and Franklin D. Roosevelt Jr., which was held at the bride's family home, Owls Nest, just a stone's throw from Kennett Pike.

In the 1950s and 1960s, many of the 19th-century buildings and homes in the village were adapted for use as high-end retail shops. The Frederick family, longtime Centreville residents, lived along the pike and operated a butcher shop and later converted their outbuildings into what is now Frederick's Country Center. Many local residents have frequented the local landmark, Buckley's Tavern, which is along Kennett Pike.

Over the past century, Centreville has undergone a tremendous amount of change. However, it is perhaps one of the last enclaves of open land in New Castle County. The year 1985 saw 15 buildings in the village added to the National Register of Historic Places within the Centreville Historic District.

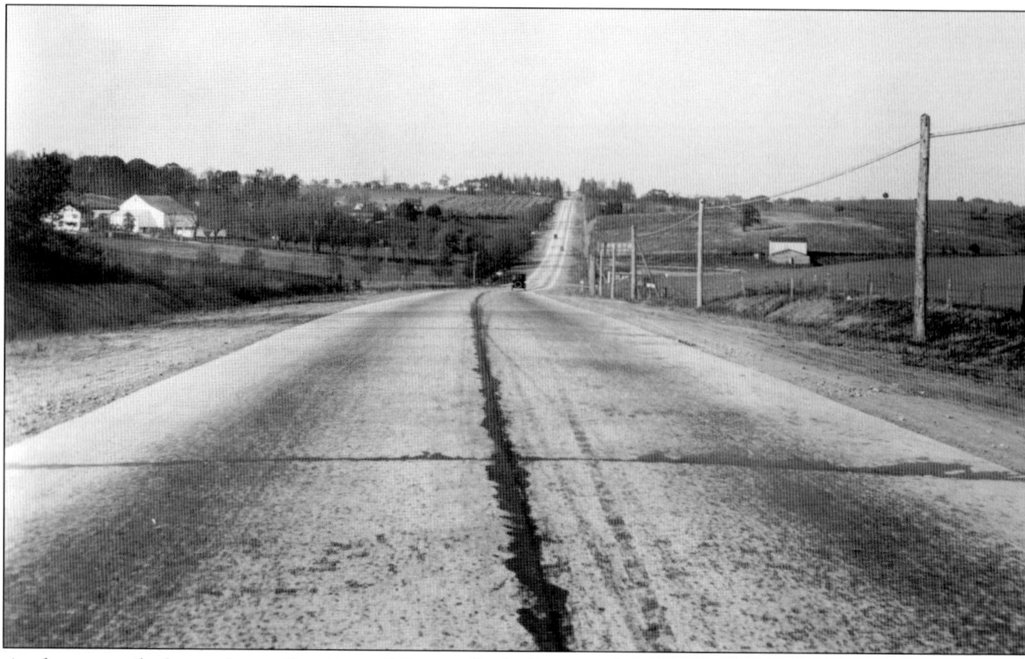

A photograph from the collection of Pierre S. du Pont, taken around 1922, shows the completed Kennett Pike looking north toward the village of Centreville, which is located at the top of the hill in the background of the image. Just north of this vicinity is approximately where P.S. du Pont commenced road construction in 1919. (Courtesy of Hagley Museum and Library.)

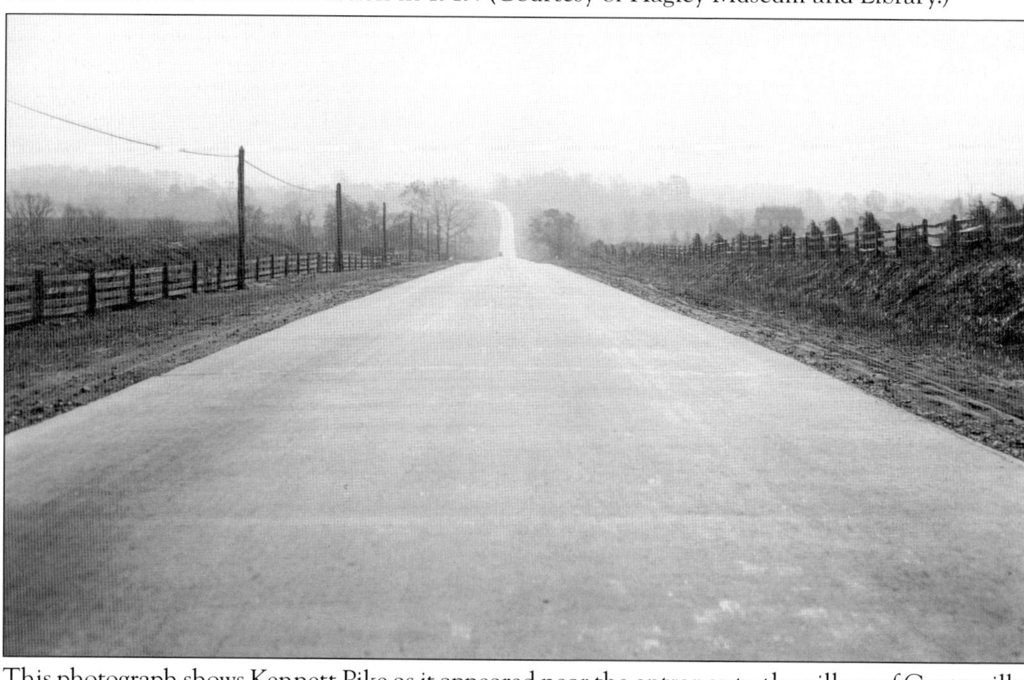

This photograph shows Kennett Pike as it appeared near the entrance to the village of Centreville. Although the view is looking away from the village and south toward the city of Wilmington, it shows the width of road at this point, as well as its rural surroundings in the early 1920s. (Courtesy of Hagley Museum and Library.)

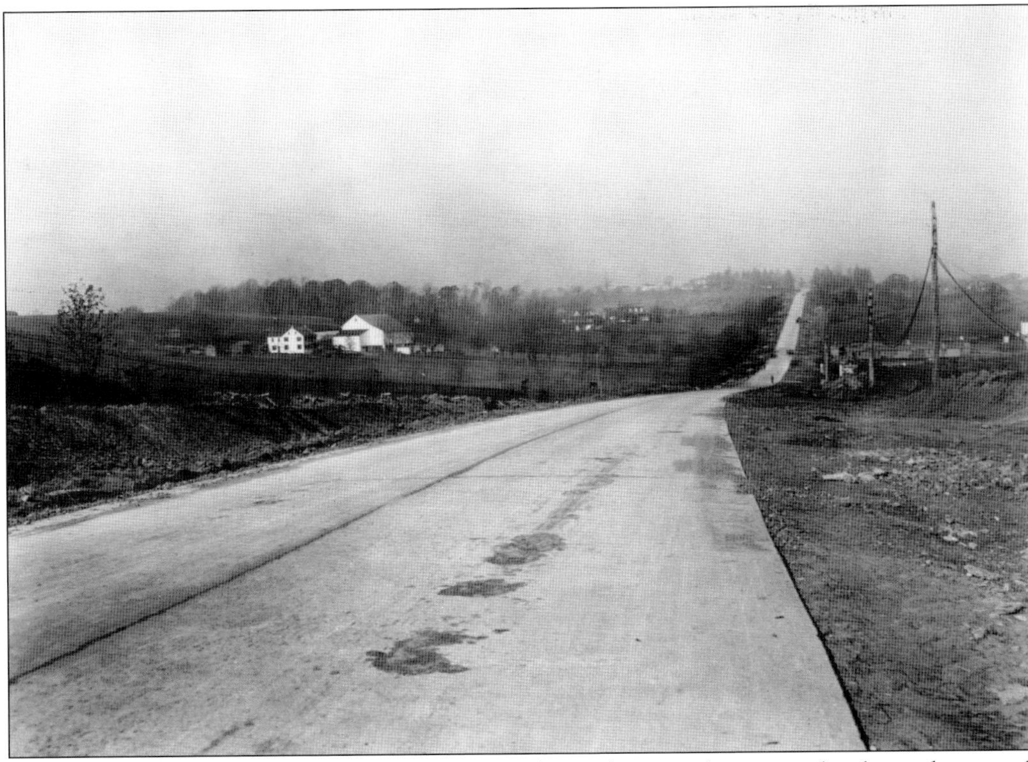

Although showing a similar view as the photographs on the preceding page, the above photograph shows Kennett Pike under construction sometime in 1919 or early 1920 with a newly paved surface and unfinished dirt shoulders. The white barn located at left still stands along the pike just north of the Lower Brandywine Presbyterian Church (shown more recently below). It is famous in the area for its display of the American flag, which is unfurled around Independence Day each year. Once a part of the W. Peters Farm, the accompanying house, which was built in 1840, still stands today with the barn. In the early part of the 20th century, travelers would stop here at a public horse trough. (Both, courtesy of Hagley Museum and Library and Edward P. Frederick.)

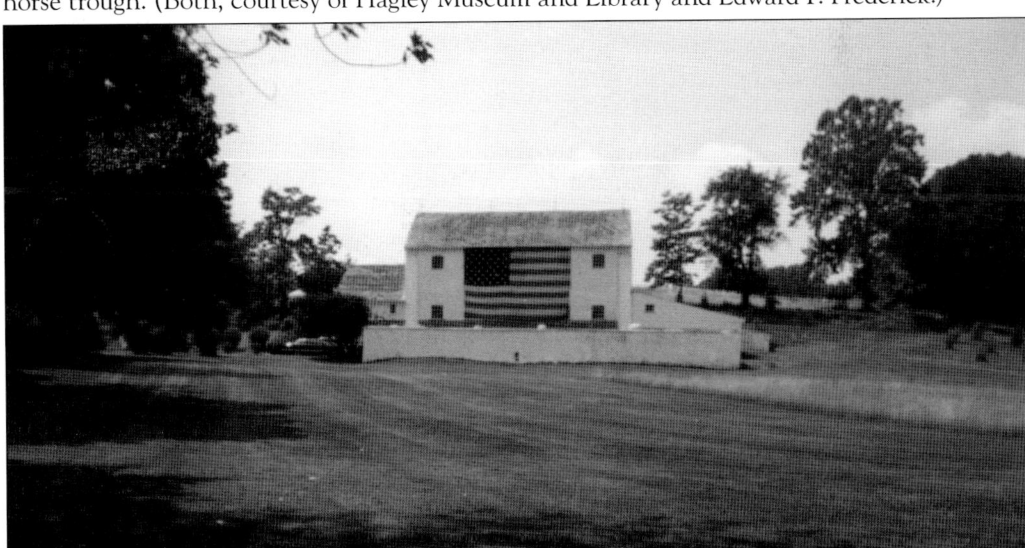

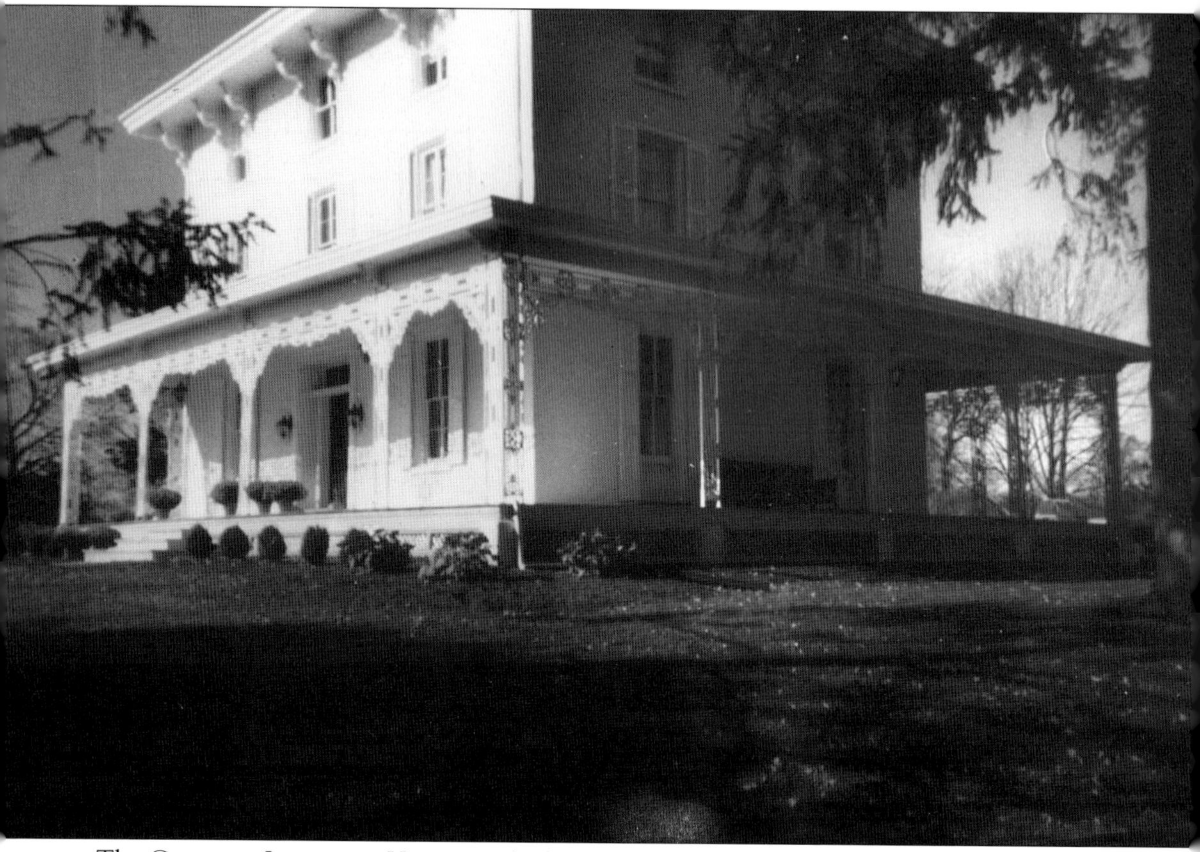

The Carpenter-Lippincott House was built in 1847 along the western edge of Kennett Pike in Centreville. Originally called Riverview, the house was located at the highest point of the nearly 200-acre property, which at one time lay along both sides of the pike. The Carpenter family lived in the house until the beginning of the 20th century, when it was bought by Clement Lippincott. Lippincott was the owner of a dry goods store in downtown Wilmington on Market Street. In the late 1990s, the property was subdivided into three separate parcels. The original house has since been meticulously renovated. (Courtesy of Edward P. Frederick.)

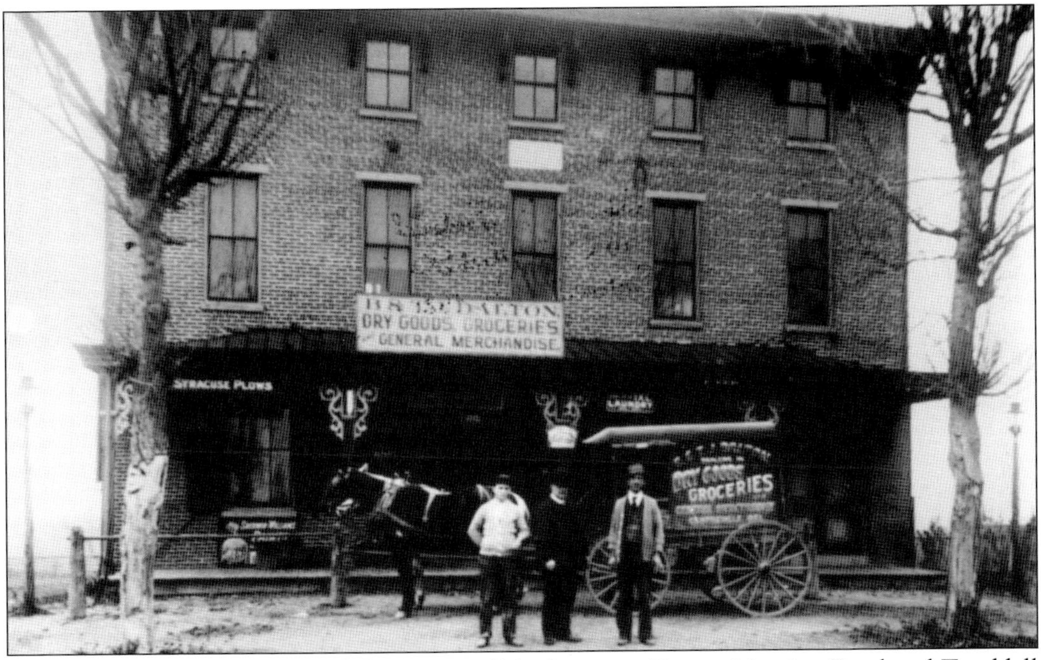

Located on the eastern edge of the Kennett Pike between Center Meeting Road and Twaddell Mill Road was Dalton's Store, shown above in the early 20th century and below in the 1970s. The building itself was constructed in 1876 as Centreville Lodge No. 37, Independent Order of Odd Fellows (IOOF) after an outcry in the community for a public gathering space. Thomas Dalton moved his store to this location in 1877. In 1921, W. Frank Smith, who had worked for Dalton since 1922, bought the business, which remained in operation until 1966. Today, the Odd Fellows continue to meet on the third floor, and the first floor is occupied by Wild Thyme flower and gift shop. (Both, courtesy of Edward P. Frederick.)

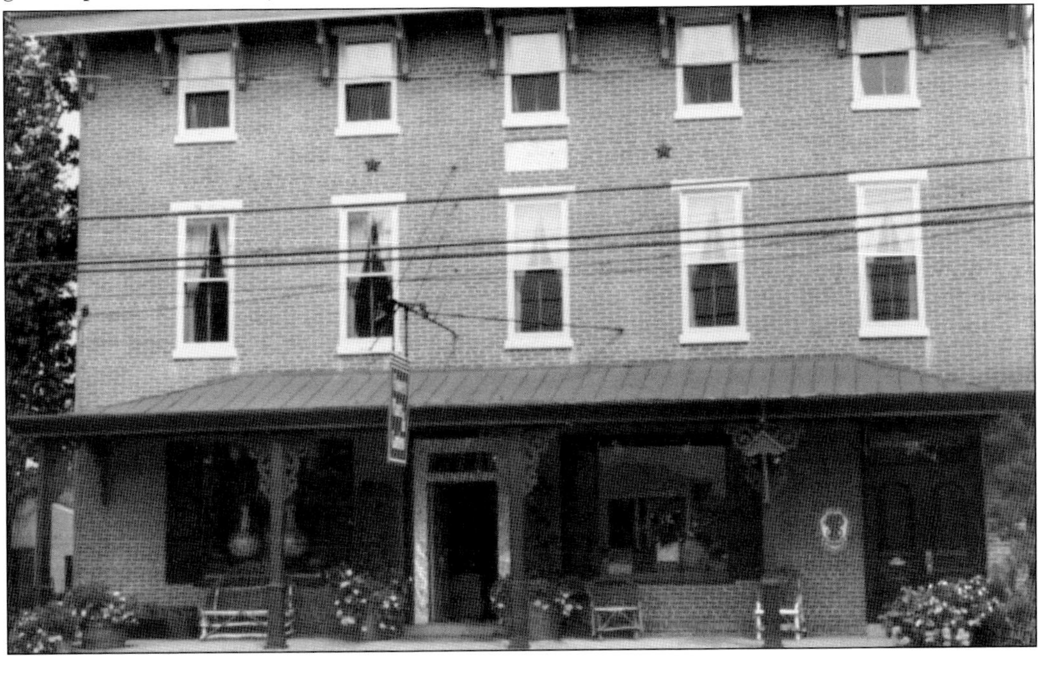

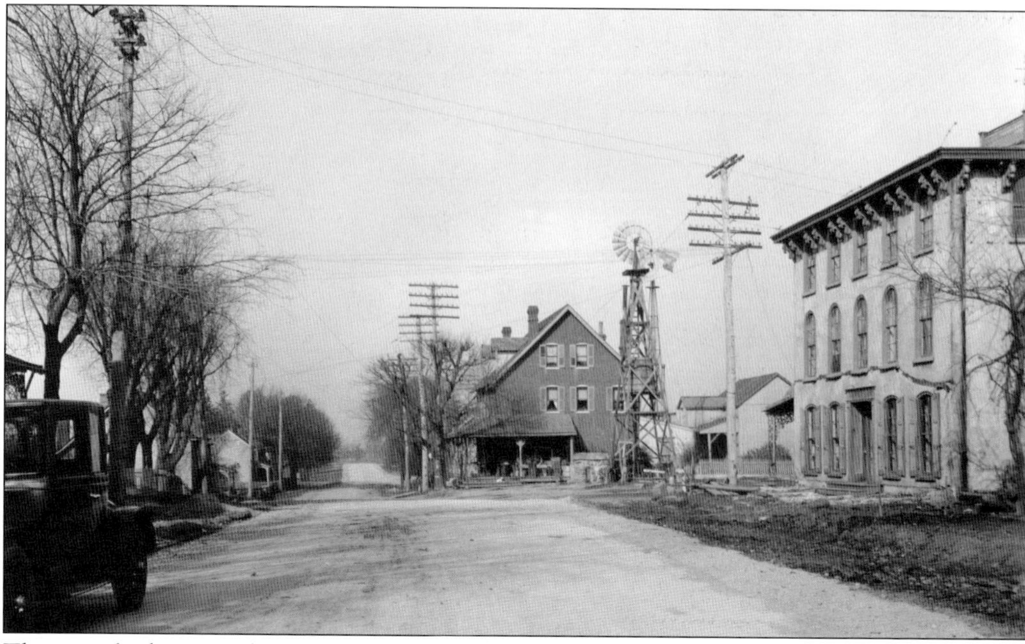

This view looking north through the village of Centreville shows Kennett Pike as it would have appeared during its renovation between 1919 and 1920. At right is the Todd Hotel, and farther up the road is Conner's Store. Twaddell Mill Road runs between the two buildings. (Courtesy of Hagley Museum and Library.)

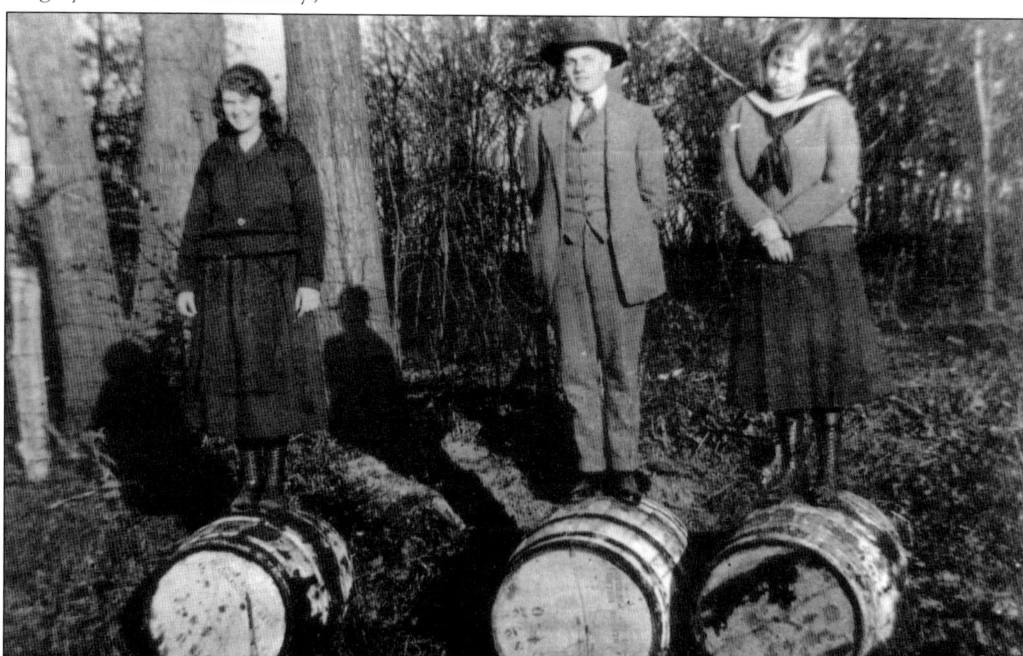

Pictured from left to right, Centreville residents Elizabeth Gregg, Ed Frederick Sr., and Marion Dalton stand atop whiskey barrels from the Todd Hotel on November 19, 1919. All three teenagers were from longtime Centreville families. The Greggs came to the area near Centreville around 1840, the Daltons, in the early 1800s, and the Fredericks, in 1900. (Courtesy of Edward P. Frederick.)

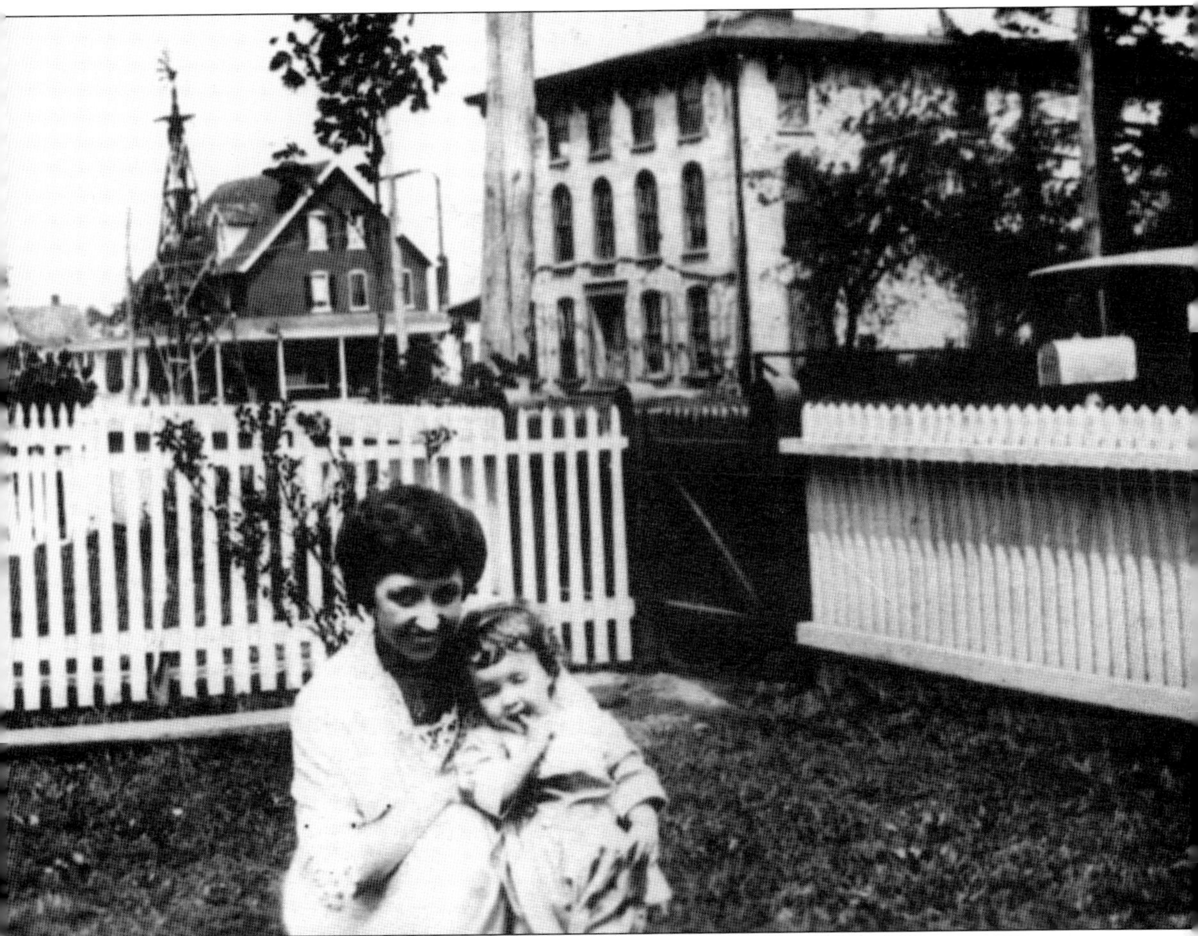

The Todd Hotel, pictured here around 1920 (behind members of the Frederick family), once stood at the site of the present-day Canby Grove Park. The hotel was a popular stopping point for travelers during the 19th and early 20th centuries. The original building on the site, the Todd Inn, was demolished in the 1890s and replaced with the Todd Hotel. Many of the laborers who worked on the widening of the pike frequented this establishment. It was torn down in 1920 to make way for the widening of the road. When demolition was complete, workers found a Hessian helmet on the site. The land was later donated by Wilmington Trust (1926) to the City of Wilmington for use as a park. It is named in honor of Henry Canby. The gazebo at the site was recently dedicated in memory of local philanthropist Rosa Laird Hayward McDonald (1916–2009). (Courtesy of Edward P. Frederick.)

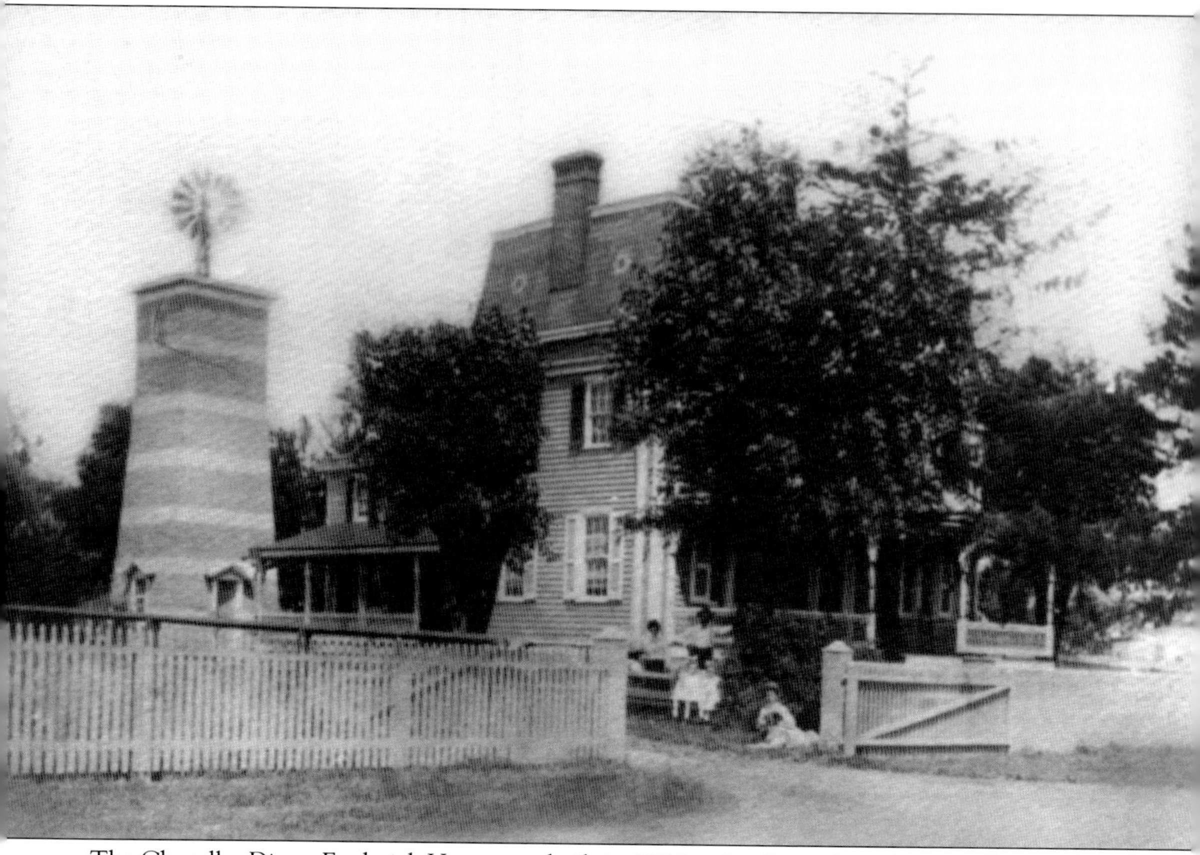

The Chandler-Dixon Frederick House was built in 1880 and still stands in Centreville along the western border of the pike near the intersection with Owls Nest Road. Three generations of the Frederick family have lived in the house since its purchase in 1900 by John Frederick. At one point, three generations of the Frederick family owned and operated a slaughterhouse and butcher shop behind the house. In the mid-1950s, John Frederick's grandson Edward P. Frederick began to convert the outbuildings into what is now known as Frederick's Country Center. Fondly known as "Ed," he is the unofficial mayor of the village. (Courtesy of Edward P. Frederick.)

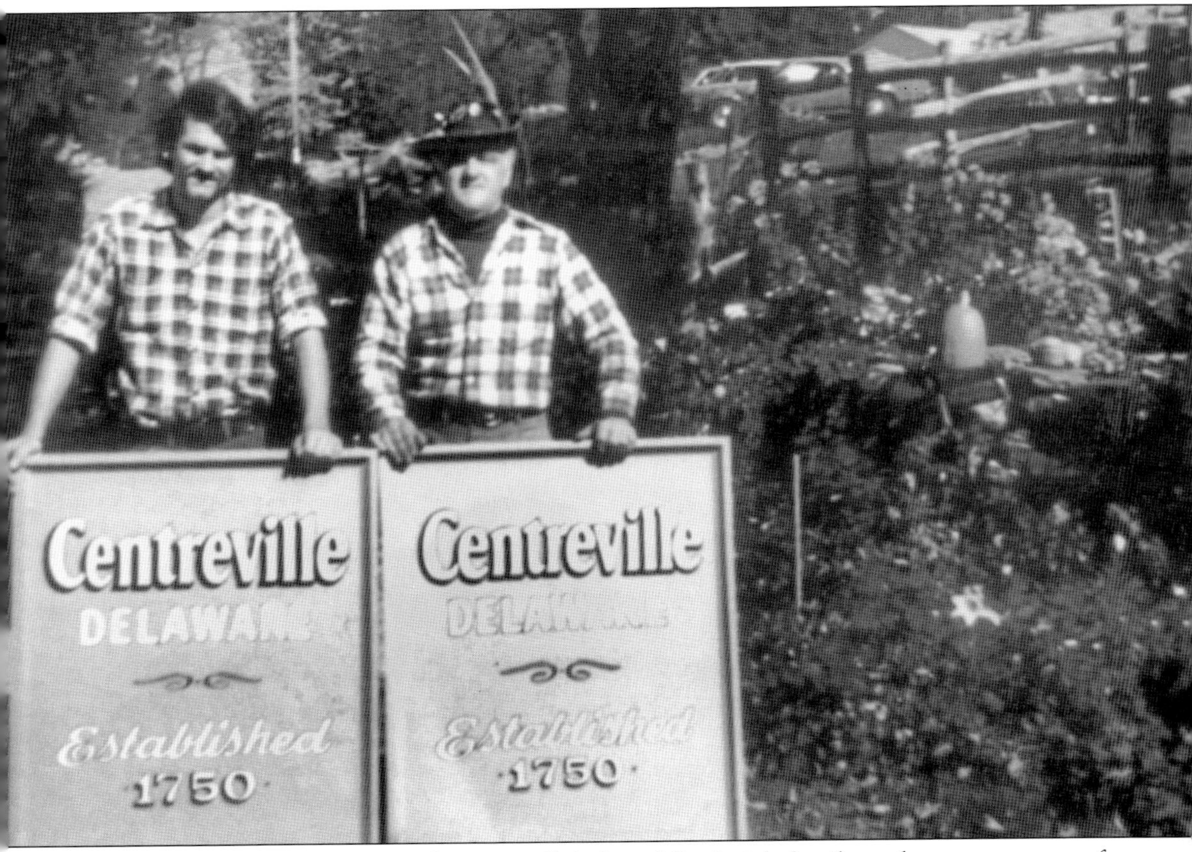

Edward P. Frederick, member of Centreville's famed Frederick family and present owner of Frederick's Country Center, at one time owned 50 Cadillacs that were parked on the family property in Centreville. Ed (pictured with his son Edward Frederick II) has since sold his collection of Cadillacs and is presently the owner of a 1945 Army jeep. He started his career in the family butcher business but later transformed its outbuildings into retail shops that are still in operation today. For many years, Frederick cut the grass and raised the American flag every day at Canby Grove Park, which is just across the street from his home and business. A local legend, he is also the go-to source for all things related to the history of Centreville and the surrounding area. (Courtesy of Edward P. and Marian Frederick.)

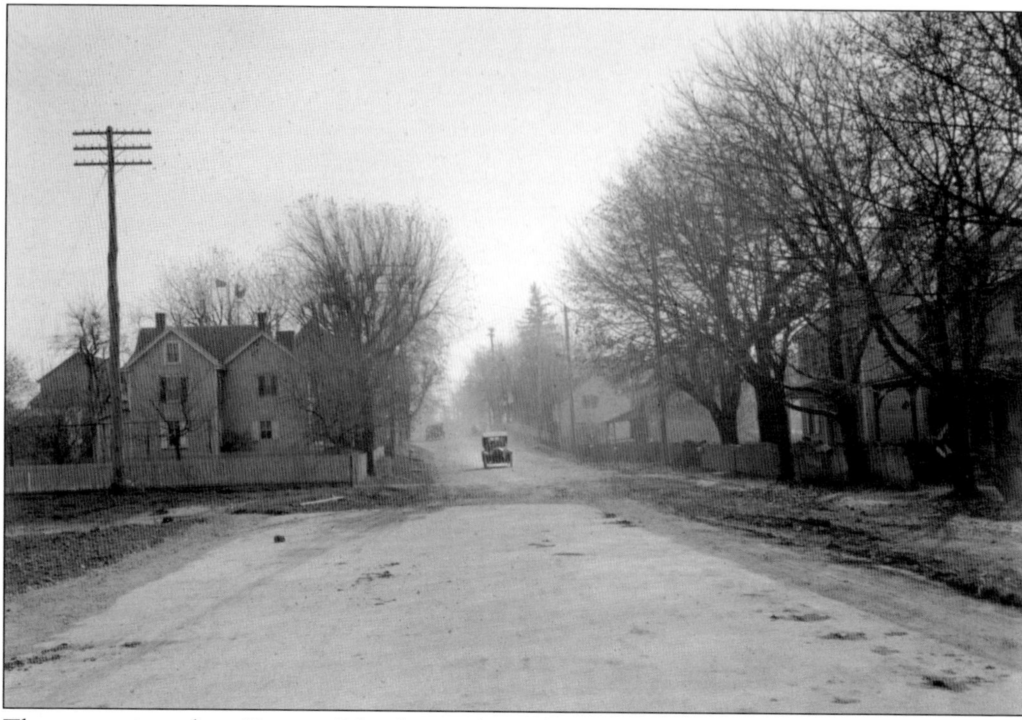

These two views show Kennett Pike during the widening of the road by Pierre S. du Pont around 1920. Both images are taken from approximately the same location, looking south through the village of Centreville. The image above shows the unpaved surface that drivers navigated before the construction project commenced in 1919. In the image below, one can see the width of the newly paved surface versus the old road before the house at left was either moved or demolished. To the right is the approximate present-day site of Buckley's Tavern. The garage at right was Centreville's only filling station. (Both, courtesy of Hagley Museum and Library.)

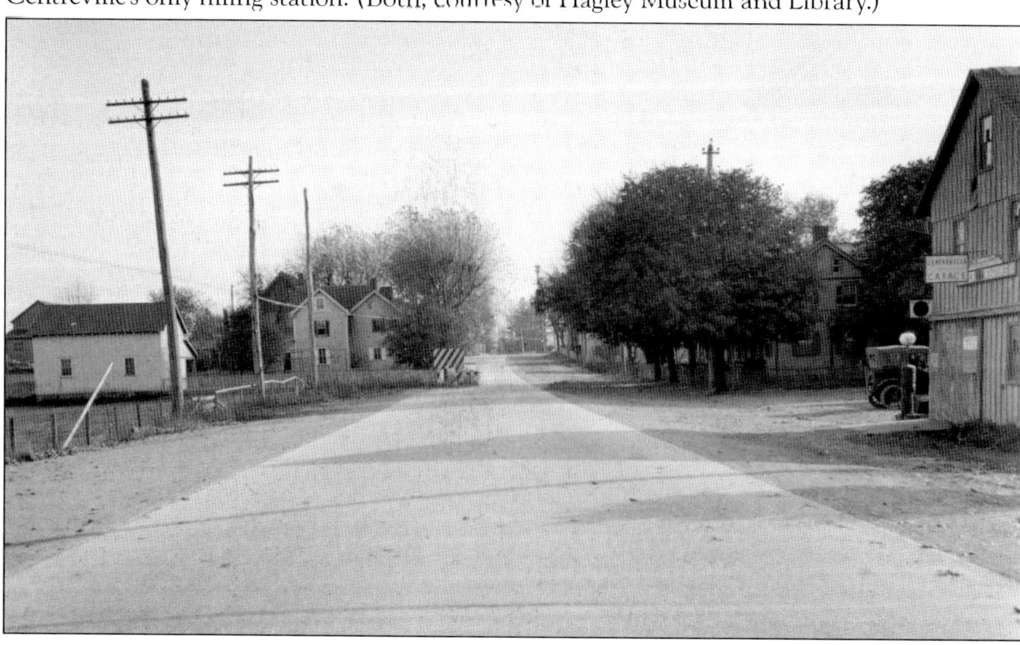

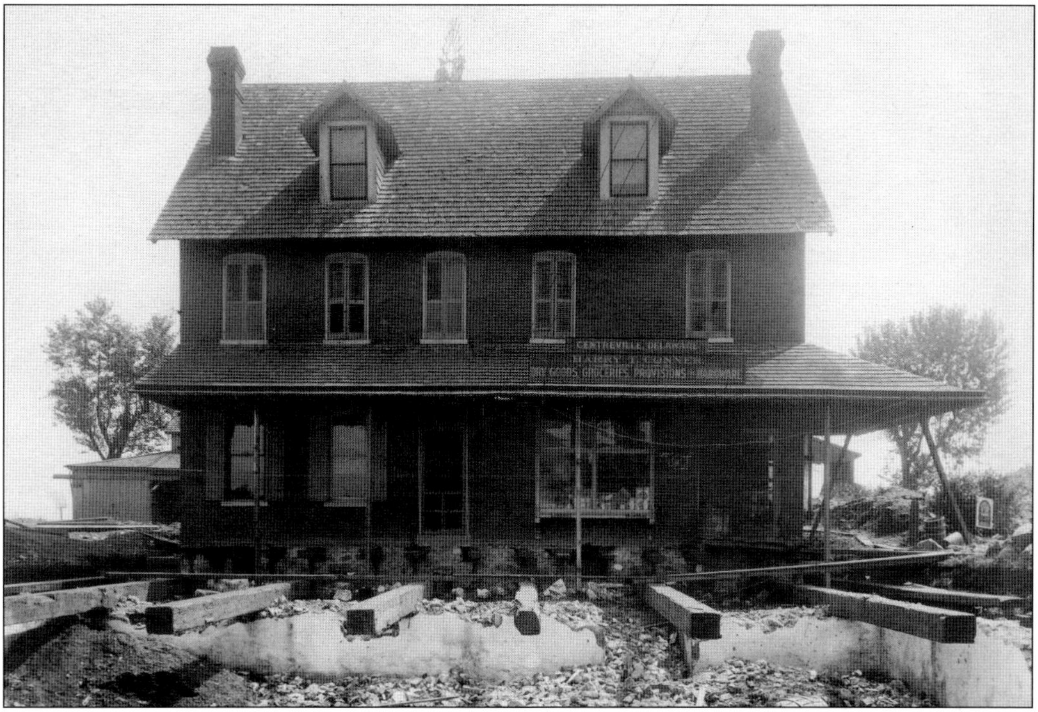

Pictured here is what was known as Conner's Store in 1920 when it was raised and moved to make way for the new and improved Kennett Pike. Located on the pike's eastern edge, Conner's was built in 1900 by Harry Conner. Previously, this had been the site of the A.S. Smith Hotel and the Centreville Tavern, which was built in the early 19th century. Until the post office began rural delivery, this dwelling was also Centreville's post office (1900–1948). Just opposite Twaddell Mill Road from Canby Grove Park, Conner's was operated by Harry Conner's sons Percy and Norman after his death in 1948. The sons ran the store until 1978. (Both, courtesy of the Longwood Library.)

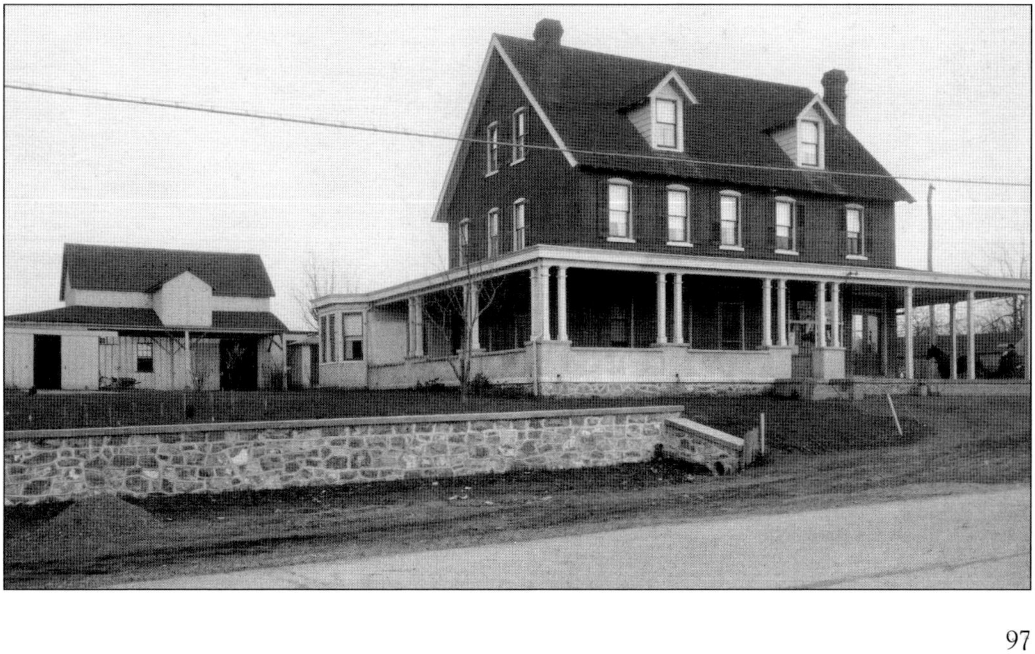

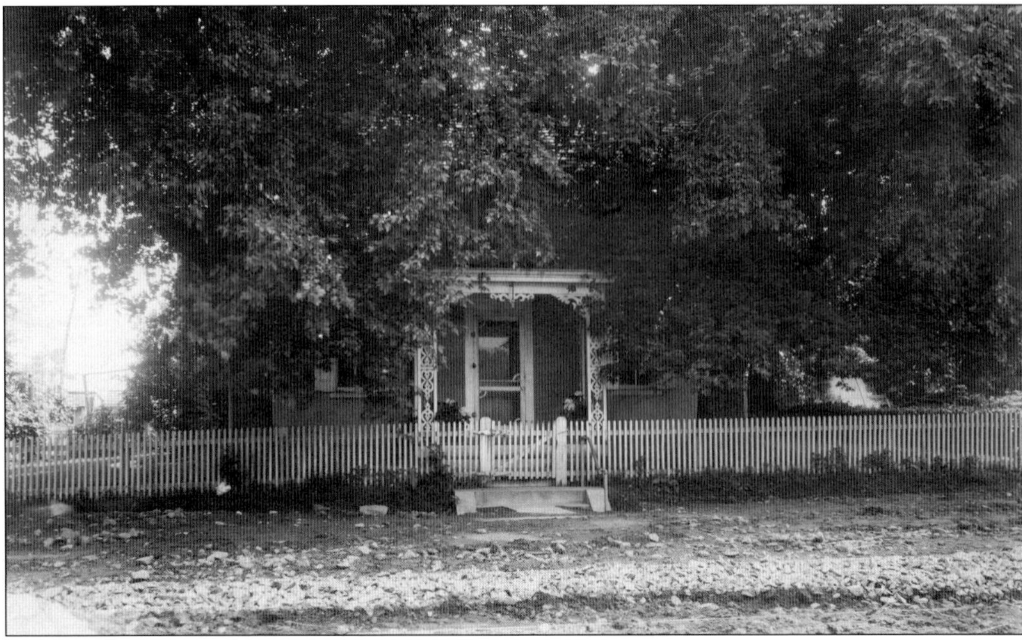

Both images on this page show the construction on Kennett Pike in 1920. The number of homes located closely along the pike in Centreville presented a particular challenge to Pierre S. du Pont during his project. Du Pont and his lawyers spent thousands of dollars purchasing frontage along the road and, in some instances, buying entire homes and tearing them down. At least two properties, one being Conner's store (see previous page), were moved by horse in Centreville. One other property that was moved was the Crossan home. Mr. Crossan refused to sell out to du Pont, forcing traffic to slow at this narrow section of roadway. He eventually gave in and moved his family across the street with neighbors while the dwelling was being moved. (Both, courtesy of the Longwood Library.)

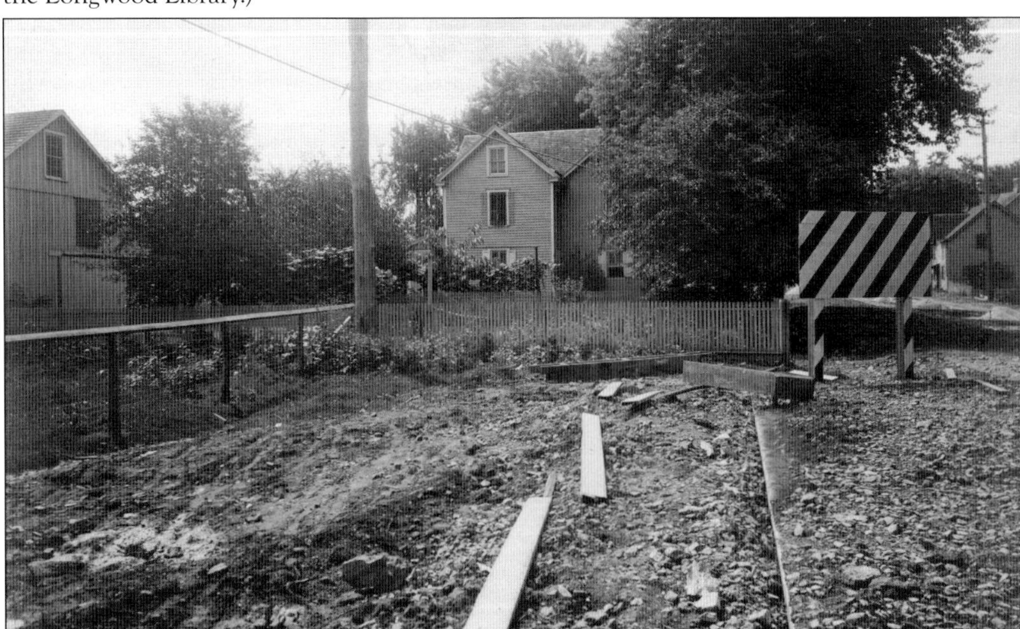

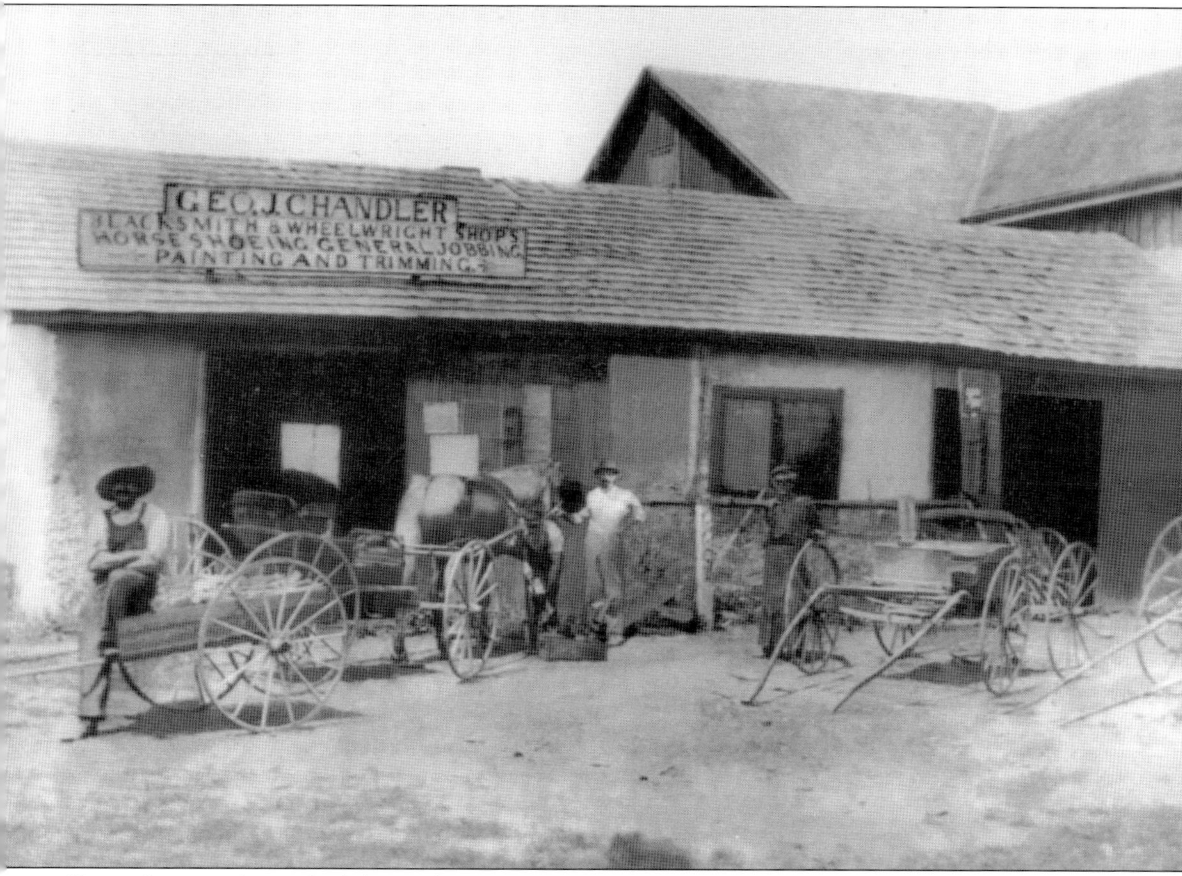

Shown here at the end of the 19th century is George Chandler's blacksmith shop, which was located at the site of what is now the parking lot of Buckley's Tavern. The blacksmith shop was situated next to his home, now the tavern. Originally owned by A.R. Nichols, it was under the ownership of George Chandler by 1879. This property was purchased in 1926 by Ted Antoine, who ran a gas and service station on the site. In addition to being in the automobile business, Antoine ran a small tavern on the first floor of what is now Buckley's. The property was purchased in 1951 by Pierre S. du Pont's nephew W.W. "Chick" Laird, who hired Dennis Buckley as proprietor of the establishment (hence its name for the past 62 years). Laird was also well known as a local philanthropist, donating a 1920s horse water trough to Centreville in the 1960s. (Courtesy of Edward P. Frederick.)

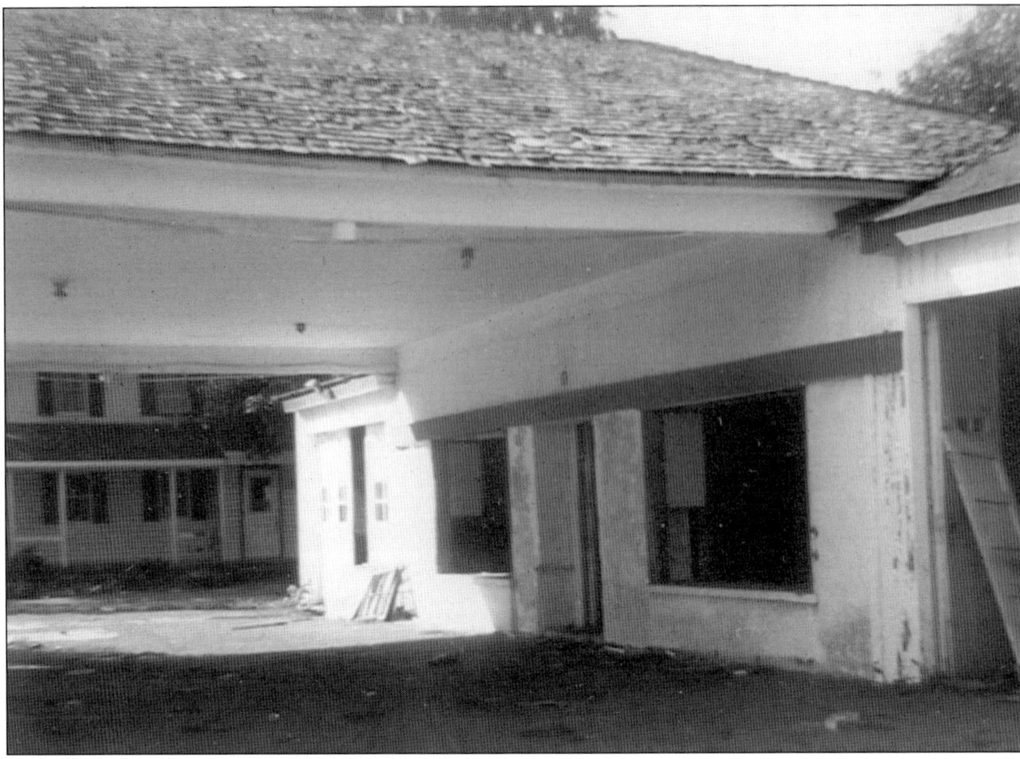

These two photographs show the vacant gas and service station of Ted Antoine in the late 1940s or early 1950s. Located on what is now the parking lot of Buckley's Tavern (which can be seen in the background of the above image), the service station had been in operation since 1926. Antoine, who lived on the second floor of the taproom next door, eventually moved to the nearby Harvey Farm, where he raised racehorses. Also on the property at one time was a small candy store, which was operated by his daughter. Ted died in 1962, and the Harvey Farm was sold to developers. (Both, courtesy of Edward P. Frederick.)

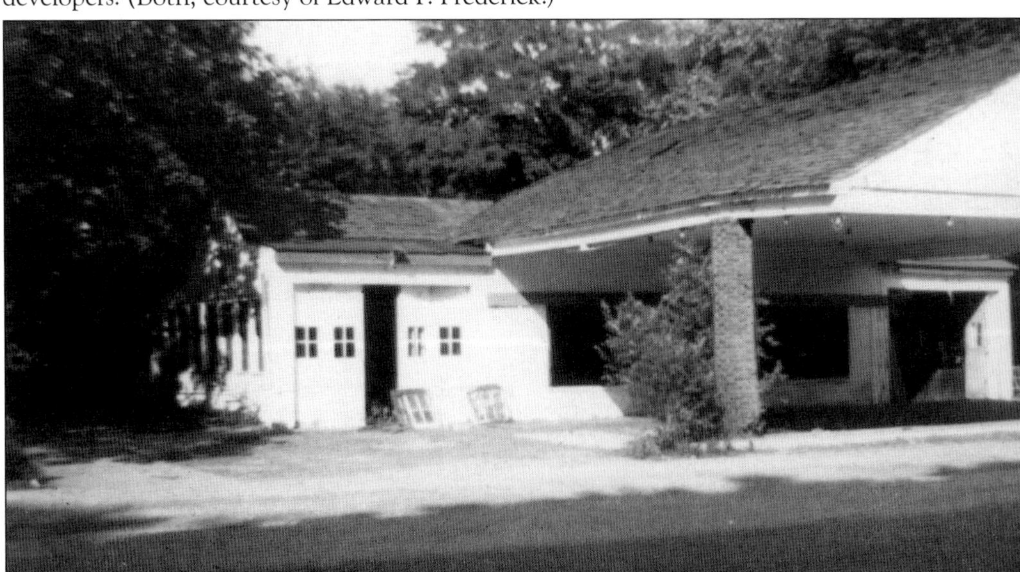

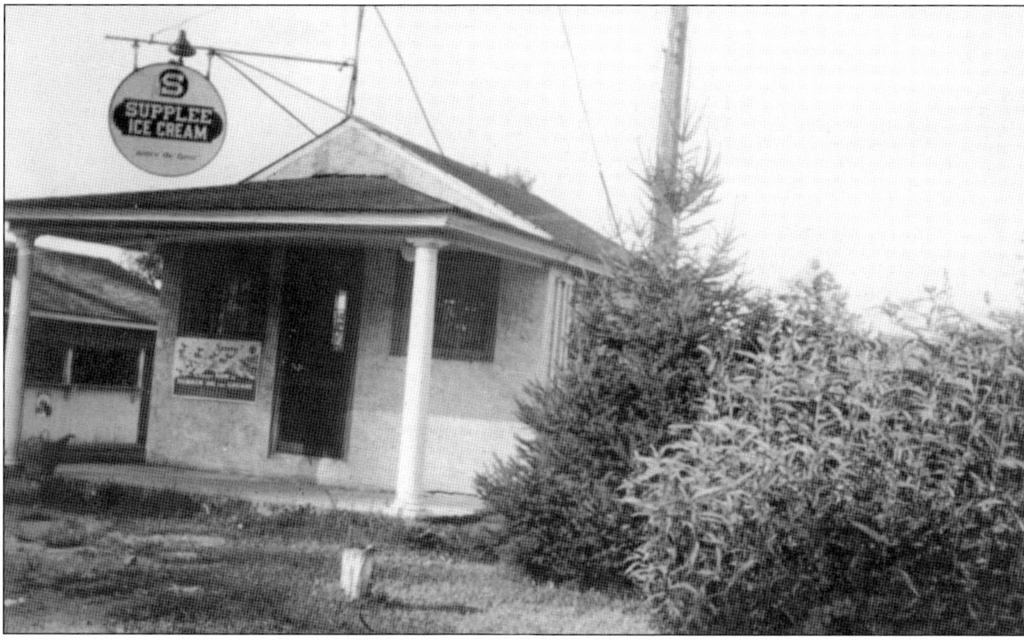

Both of the images on this page show the general vicinity of what is now known as Buckley's Tavern. The image above shows the ice-cream and candy store operated by Ted Antoine's daughter just past her father's gas station along Kennett Pike. In the image below, viewers can see Kennett Pike heading toward Wilmington in 1931. At right is a small store followed by the awning of Antoine's gas station, which sits on what is now the Buckley's parking lot. The young boys in the photograph are identified on the back of the image as Bill Tapie and a member of the Beattie family. (Both, courtesy of Denis O'Sullivan.)

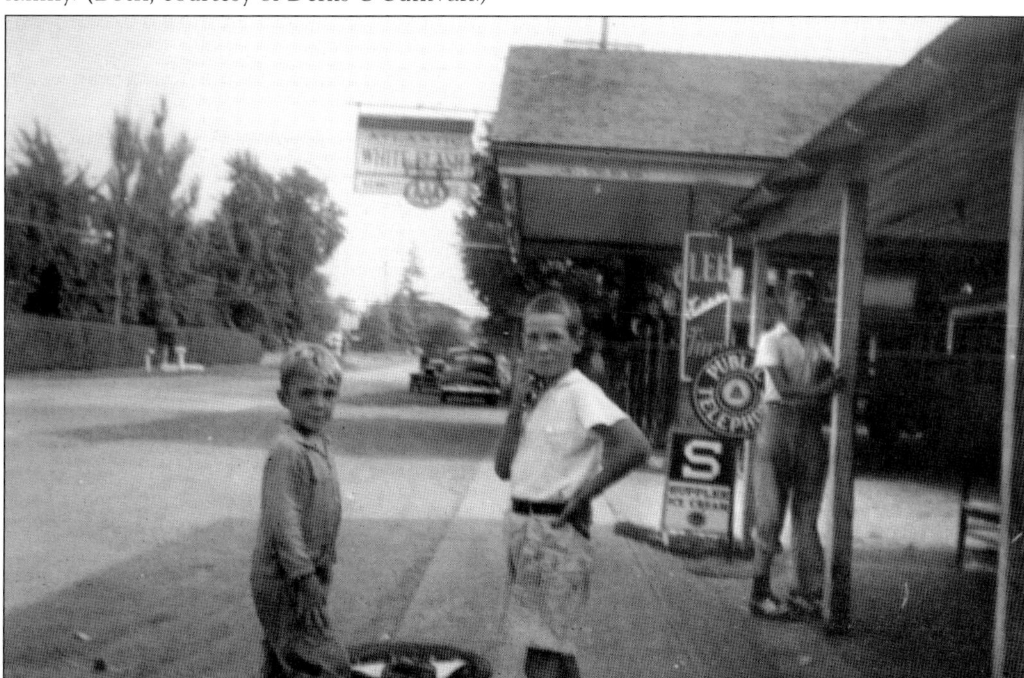

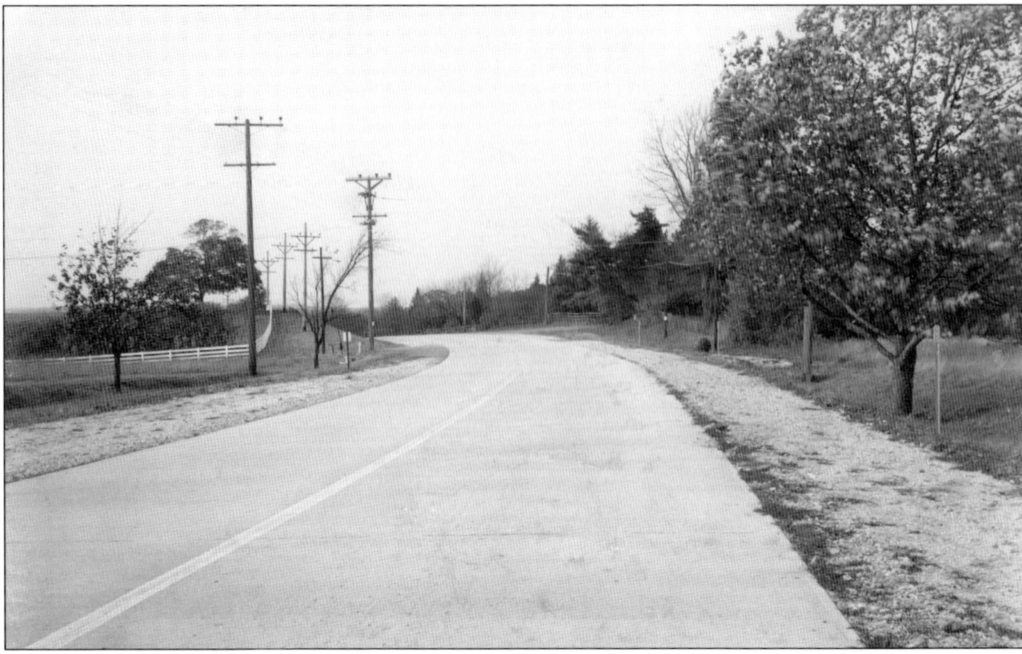

This photograph, taken in 1942, shows the improved Kennett Pike heading north toward the Delaware border with Pennsylvania. At left, just before the white fence, is the intersection with Snuff Mill Road, and at right is the entrance to the community of Selborne. (Courtesy of the Delaware Public Archives.)

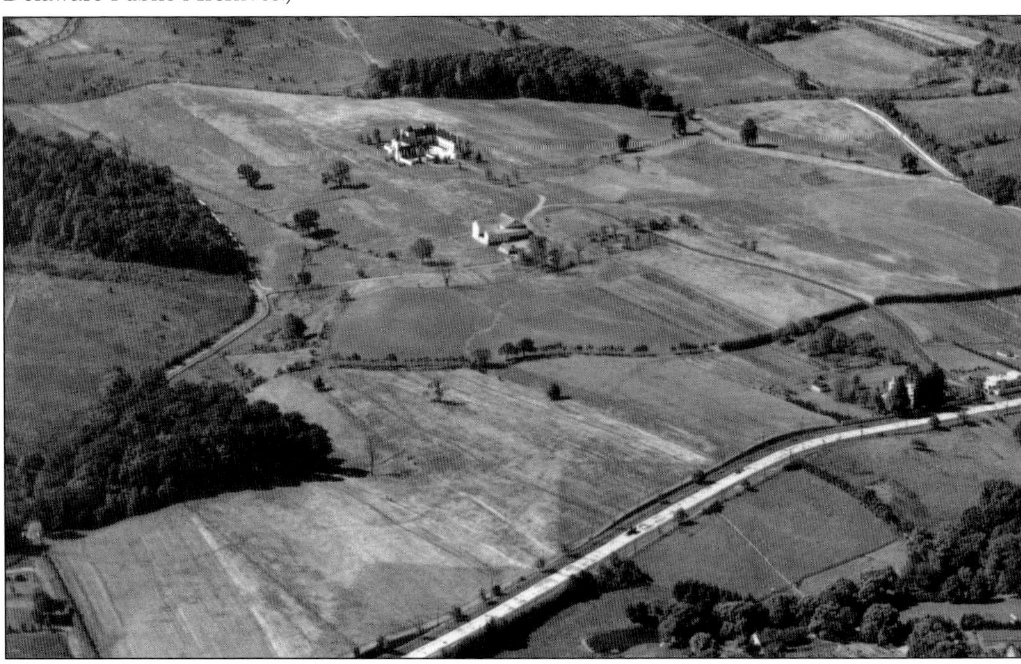

Located on the western side of Kennett Pike near the state line was Oberod, the home of Harry and Jane Lunger. The main house was built between 1935 and 1937 and was comprised of 165 acres. Following the deaths of the Lungers, the house was donated to the Episcopal Diocese of Delaware. It has since been sold and is now a private residence. (Courtesy of Hagley Museum and Library.)

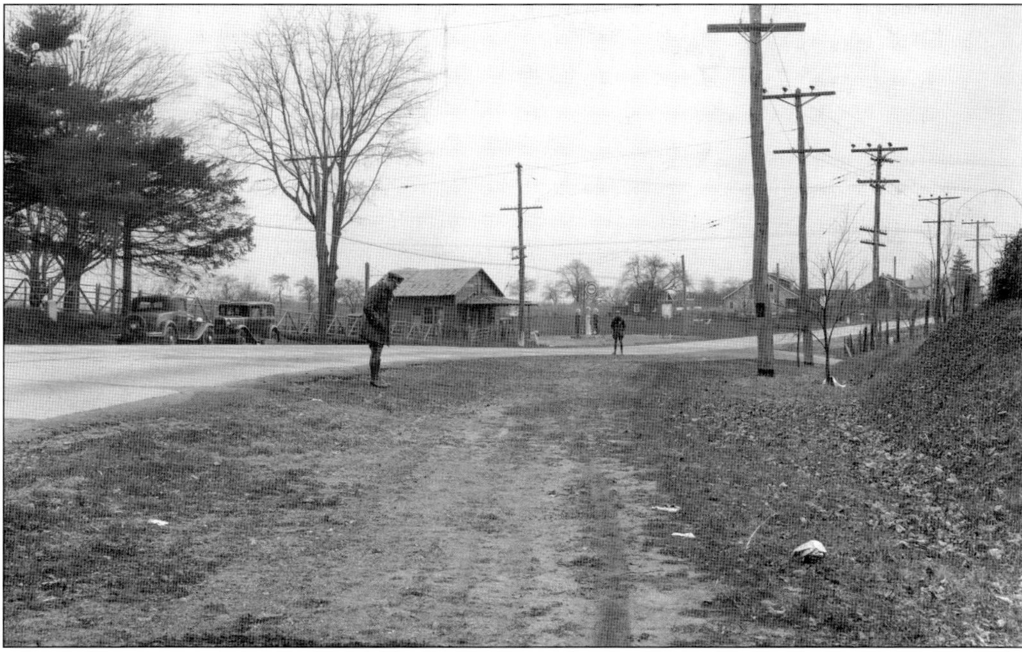

This photograph was taken as part of a fatal-accident investigation in 1932. Looking south, it shows Kennett Pike in the background as Delaware State Troopers attempt to recreate and investigate the accident that occurred here and killed two men just north of Snuff Mill Road. (Courtesy of the Delaware Public Archives.)

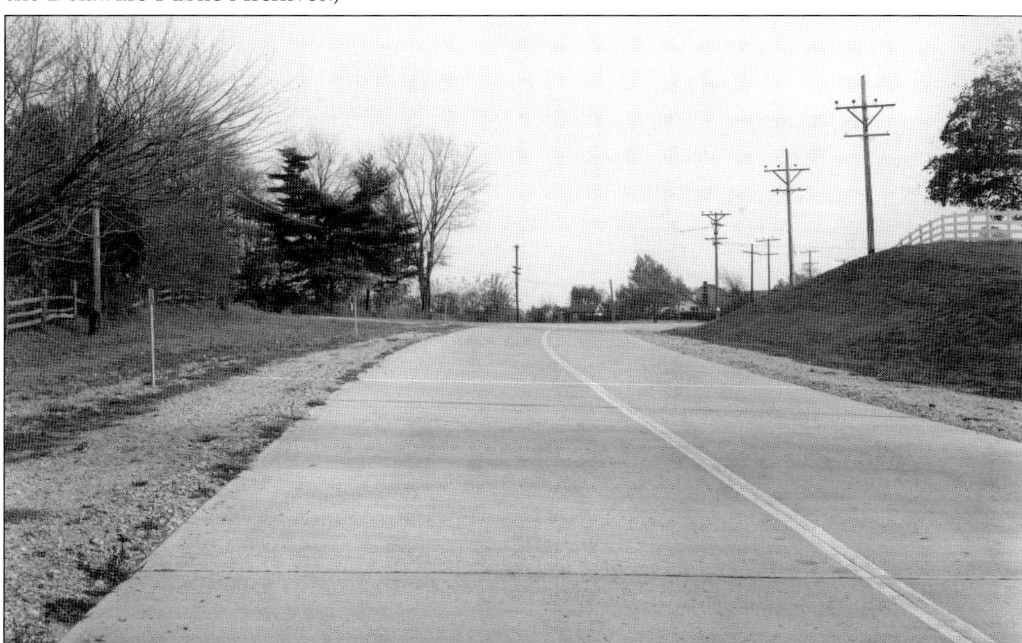

This photograph shows Kennett Pike looking south from the state line into the village of Centreville in 1942. The modern-day community of Selborne is located at left, and Oberod, the estate of Mr. and Mrs. Harry Lunger, is located at right. More than 20 years had passed by this point since the widening project undertaken by Pierre S. du Pont. (Courtesy of the Delaware Public Archives.)

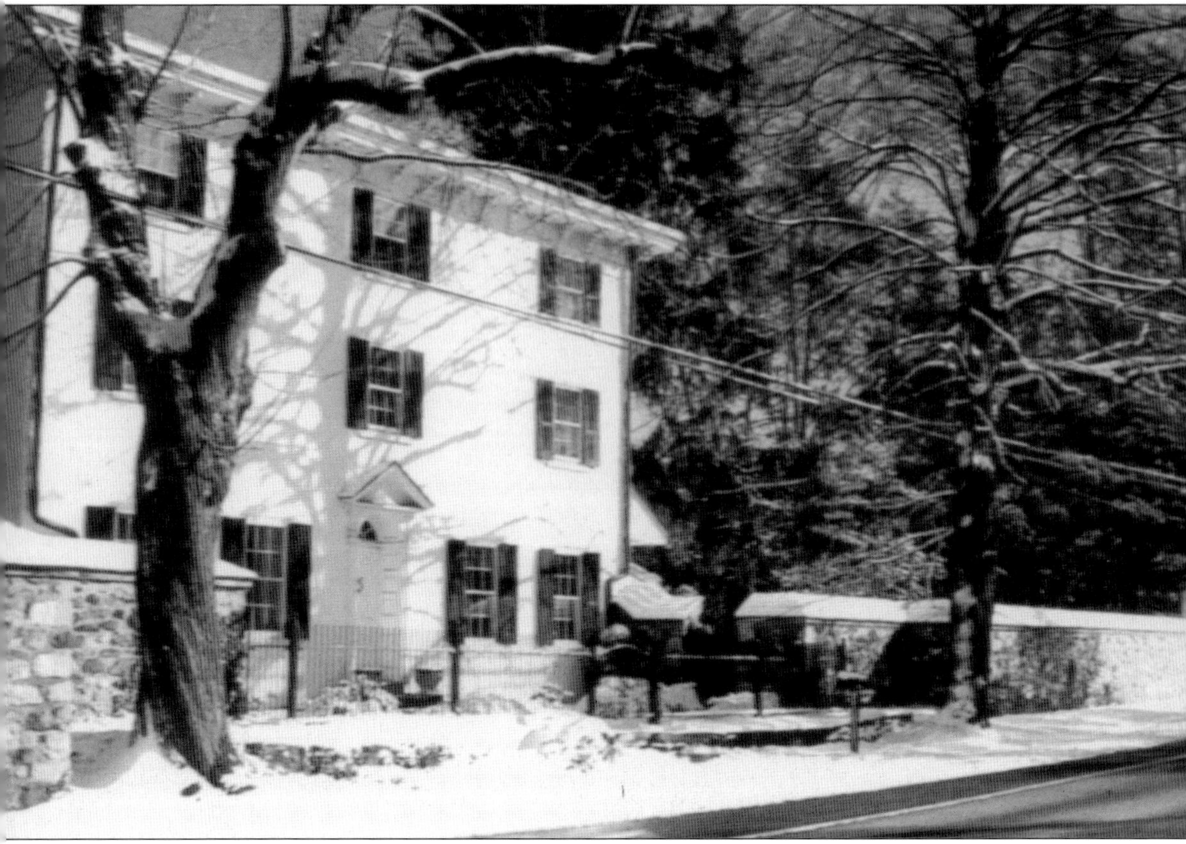

Pictured here is the Spread Eagle Tavern, also known as the "Line House." Located just north of the village on the eastern side of the pike, the dwelling was built around 1814 at the intersection of the old and new arc line that was drawn as the border between Delaware and Pennsylvania. The first line was surveyed and drawn in 1701 and was drawn again in 1764 by Mason and Dixon. The dwelling was built originally as a tavern by Charles Twaddell and served as a gathering place until the beginning of the 20th century, when it was converted to a private residence. Travelers passing by can still see the vintage concrete pillar that marks the border between the two states. It was not included in the original Centreville National Historic District as designated by the National Park Service. (Courtesy of Edward. P. Frederick.)

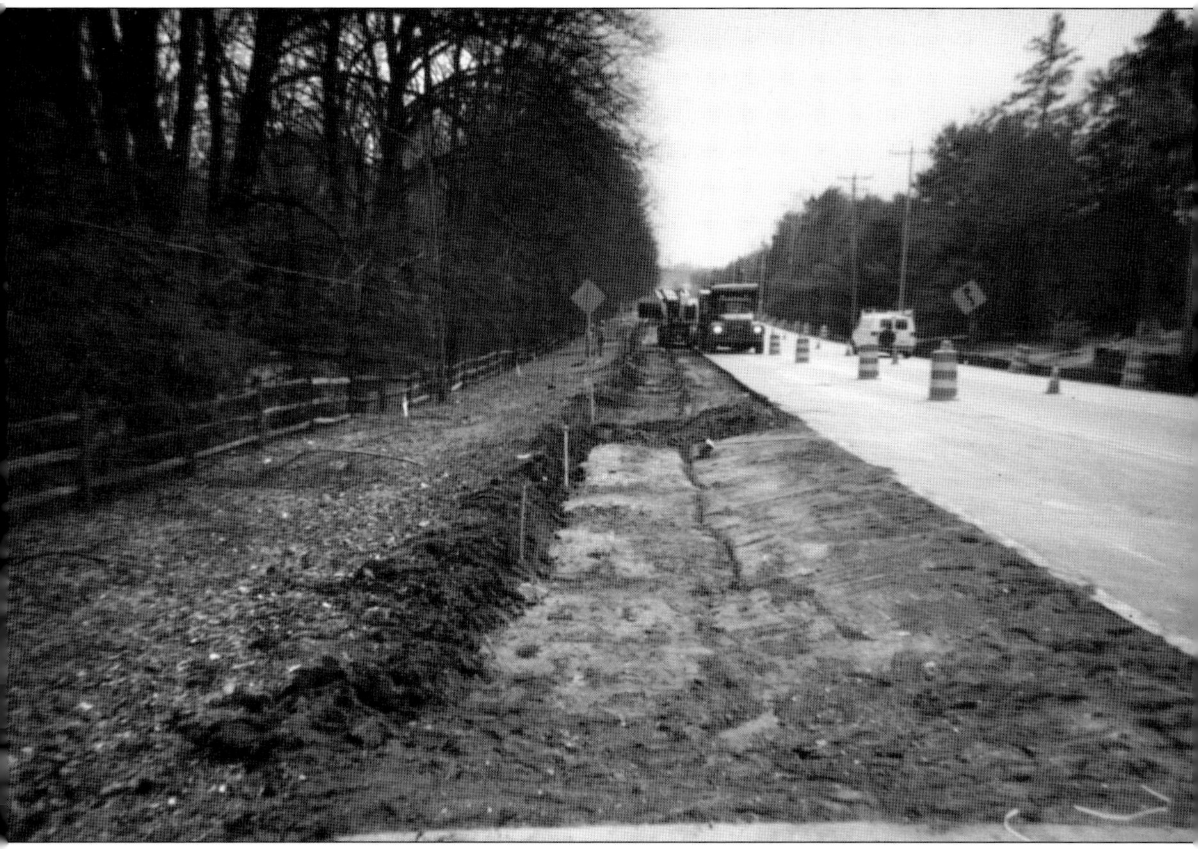

Since the widening of Kennett Pike by Pierre S. du Pont, which took place in the second decade of the 20th century, the pike has undergone many more construction projects. One of the most recent in Centreville was the addition of large planters at intersections and, in the past year, the addition of new sidewalks. This photograph taken in the late 1990s shows one of the most recent projects in progress. (Courtesy of Edward P. Frederick.)

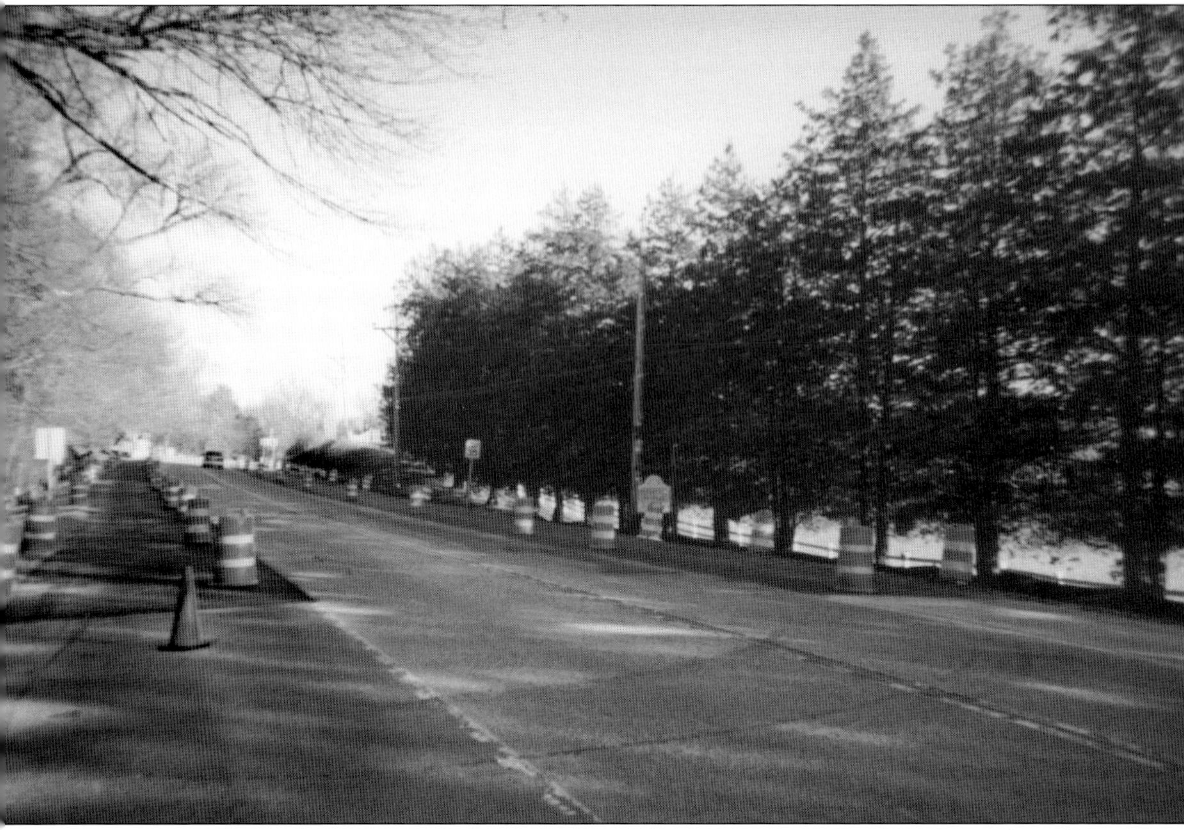

This photograph, also taken in the late 1990s, shows the addition of sidewalks to the village. (Courtesy of Edward P. Frederick.)

Five

Chester County

Before Pierre S. du Pont began widening Kennett Pike and Pennsylvania Avenue in Delaware, du Pont widened the stretch of the road that lies in Chester County, Pennsylvania, running from the state line with Delaware to the village of Hamorton. The very first work on this portion of road began in 1917 on a piece that connected the Pennsylvania/Delaware state line to approximately Mendenhall, Pennsylvania. This 1.5-mile section of road stood as the only completed portion from 1917 until after World War I in 1919. Work on the project stopped because of the shortage of available labor as the country's young men were across the Atlantic in battle against the Central Powers.

When work resumed in May 1919, du Pont directed crews to focus on the road from Brecks Lane near Wilmington to the state line. Following the completion of that work in 1920, efforts resumed on smaller sections on the Pennsylvania portion of the road that led travelers toward Kennett Square, which is also where du Pont built his lavish country estate and garden, called Longwood. Completion of the road allowed for goods and travelers to move back and forth from Wilmington to Hamorton and to Route 1, which led to markets in Philadelphia and other points.

The Pennsylvania stretch of Kennett Pike varies little from its Delaware portion. Dotted with various estates and the villages of Fairville, Mendenhall, and Hamorton, it is perhaps this section that has remained the most rural. Where Delaware has multiple traffic lights in its five-mile stretch, the Pennsylvania section still to this day has only two lights in less than four miles of road.

Travelers today can see local landmarks such as the Fairville Inn, a local bed and breakfast, as well as the Mendenhall Inn, which features a hotel and a highly acclaimed restaurant. Just past the northern terminus of Kennett Pike is the estate of Pierre S. du Pont, Longwood Gardens, which is open to the public.

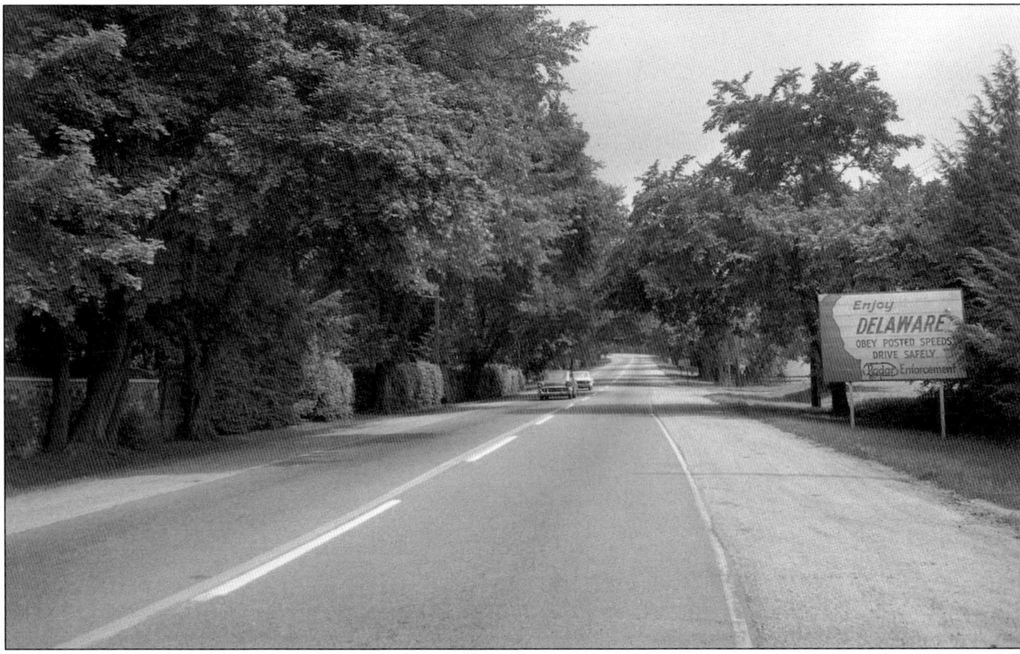

This was the view for travelers as they crossed into Delaware from Pennsylvania along Kennett Pike in 1964. At right, just before the fence, is Burnt Mill Road, and at left is the stone wall that borders the Line House just north of the village of Centreville, Delaware. This view is largely the same today. (Courtesy of the Delaware Public Archives.)

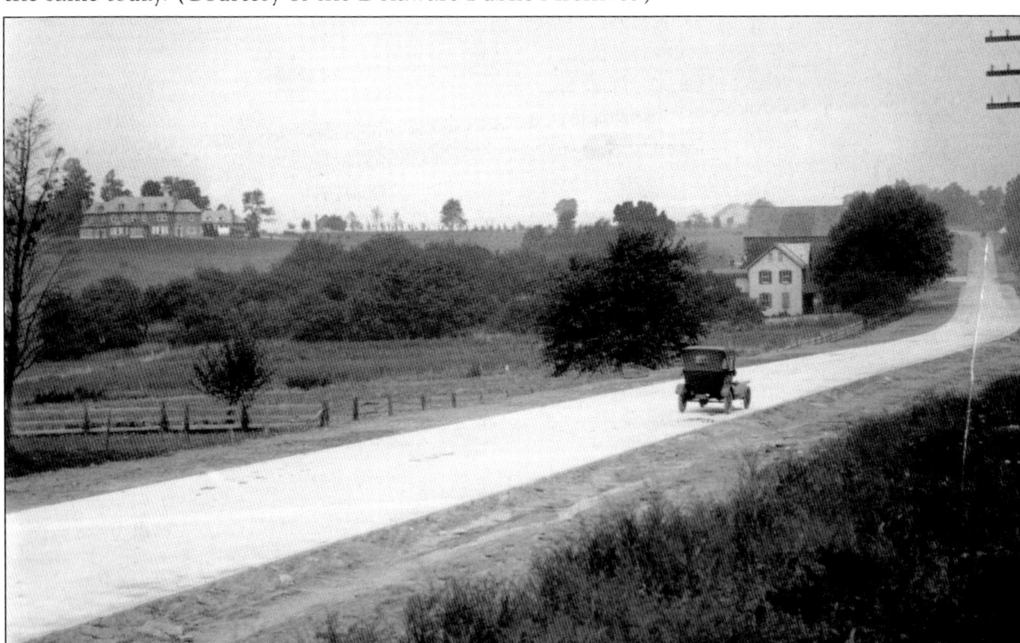

Following Kennett Pike to the north over the state line, the first village drivers and travelers would come to is Fairville, which was just north of this estate owned by Philip du Pont. This c. 1920 photograph shows the du Pont mansion at left, as well as a small farmhouse and barn, all of which still stand today. (Courtesy of Hagley Museum and Library.)

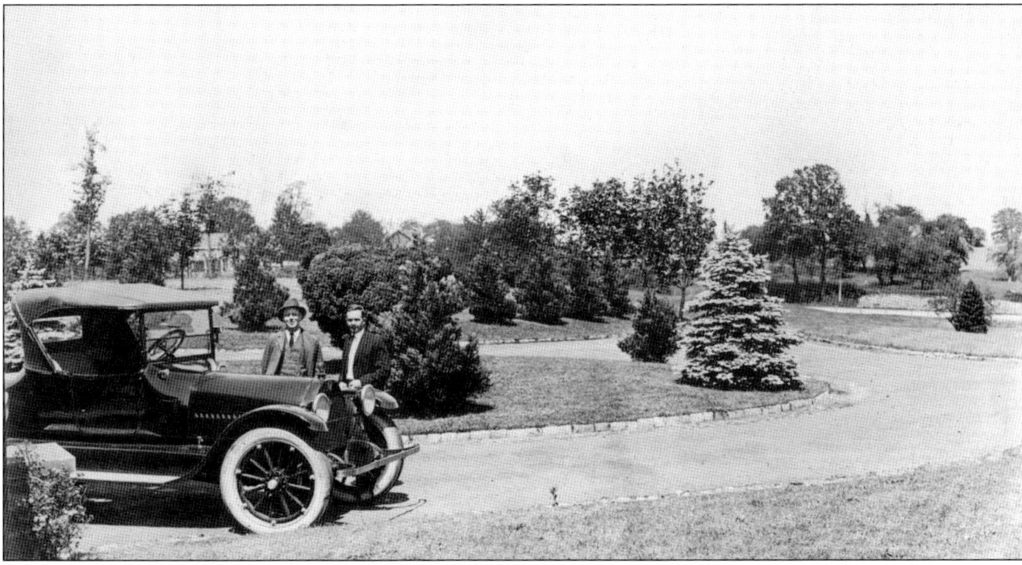

Shown here around 1920 is Philip du Pont (left) standing in the driveway of his mansion along Kennett Pike in Fairville, Pennsylvania. The other man is unidentified but may be a member of the Frederick family of Centreville. The photograph shows Kennett Pike at right, just beyond what appears to be a fence or guardrail in the background. (Courtesy of Edward P. Frederick.)

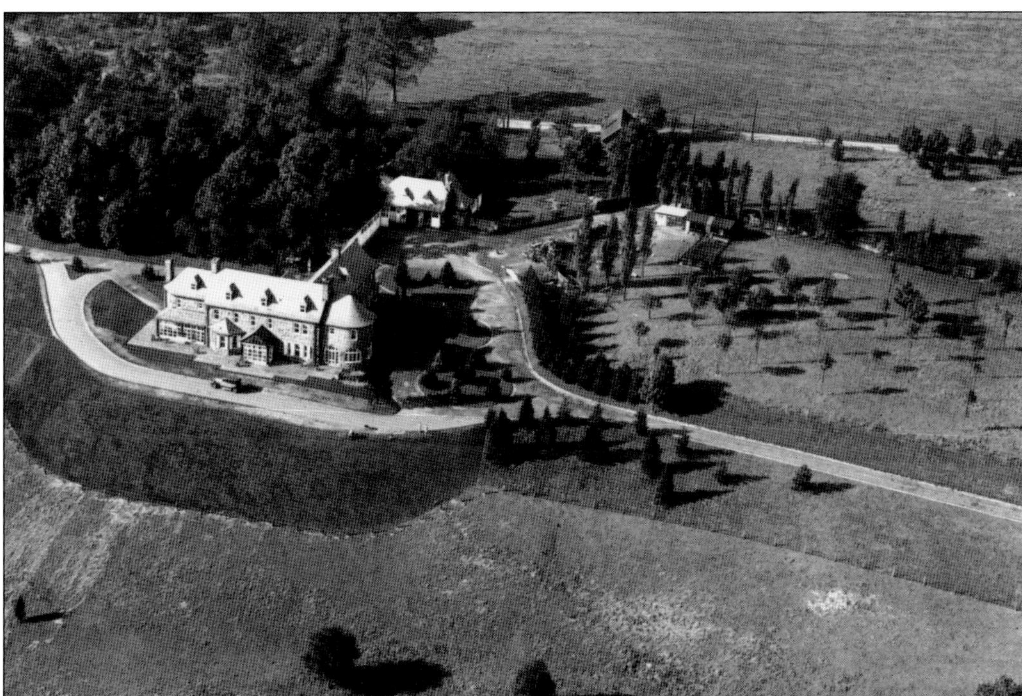

An aerial view of the mansion from 1927 shows the Philip du Pont estate a year before du Pont's death in 1928. The estate is included in the Fairville Historic District as designated by the National Register of Historic Places in 1996, which runs along both sides of Kennett Pike. (Courtesy of Hagley Museum and Library.)

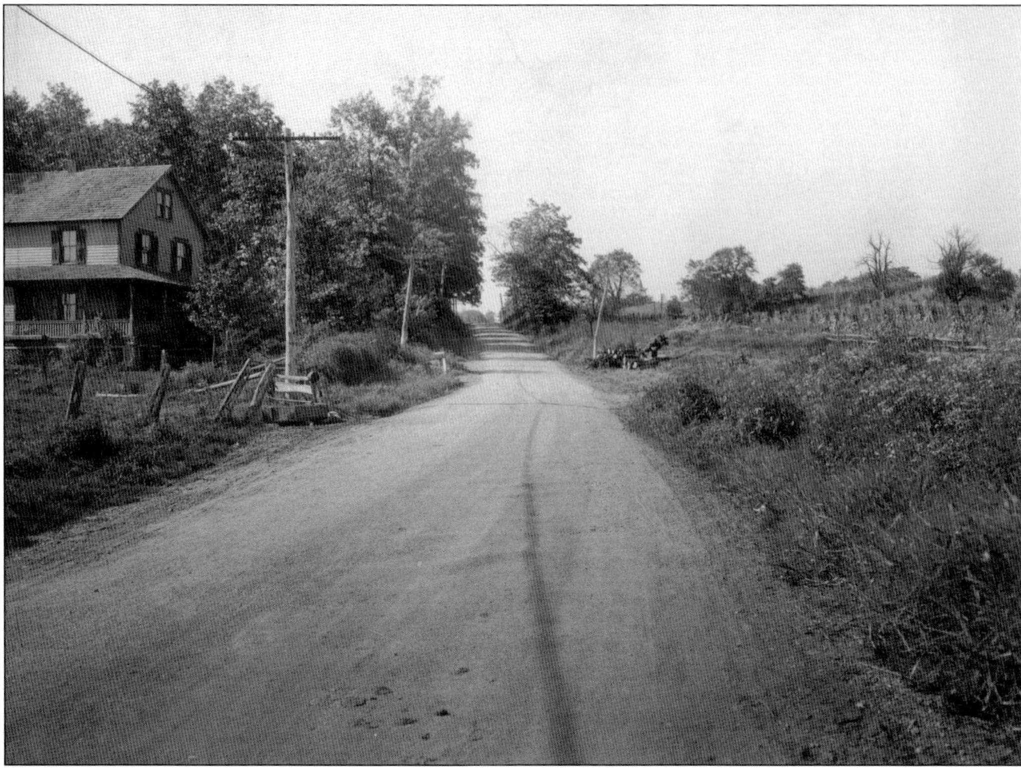

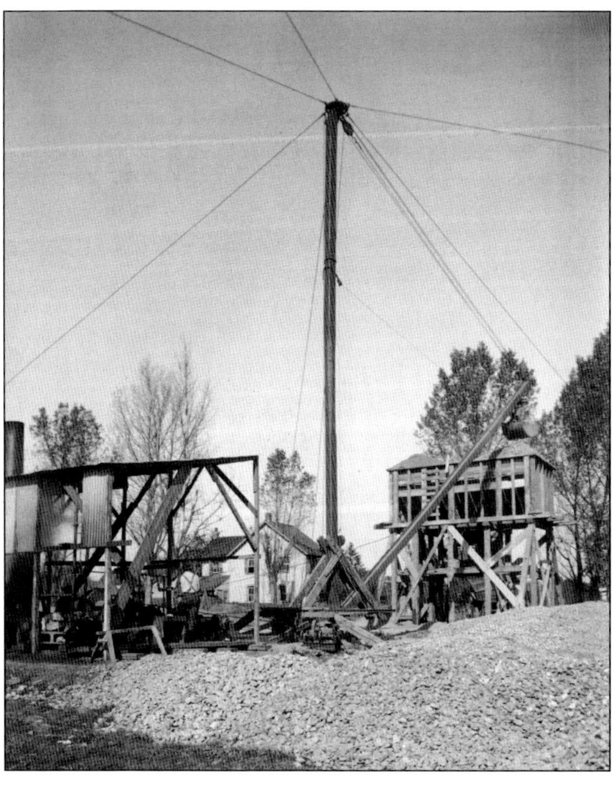

Shown in this photograph is the Kennett Pike near Mendenhall, Pennsylvania, in 1917, before the widening project undertaken by Pierre S. du Pont. This view shows the area north of the railroad tracks looking toward Hamorton. The house at left was part of the Norris Woodward farm; a water trough is located in front of the house along the road. (Courtesy of the Longwood Library.)

In the background of this photograph is the Duke Garrett House near Mendenhall, Pennsylvania; it was formerly part of the Sharpless Creamery property along the pike. In the foreground of the image is a concrete-mixing plant used during the widening of Kennett Pike around 1917. (Courtesy of the Longwood Library.)

Shown here are two additional views of the concrete-mixing plant near Mendenhall, likely taken during the first stage of the widening of Kennett Pike, prior to World War I, in 1917. The village of Mendenhall is named for Elwood Mendenhall, descendant of John and Benjamin Mendenhall, who settled in Concord, Delaware County, in 1684. The Mendenhall Inn, located in the vicinity along the eastern side of Kennett Pike, is one of the area's most popular attractions. The inn is located on land granted by William Penn to his daughter Letitia. It was first purchased by a member of the Mendenhall family in 1847. (Both, courtesy of the Longwood Library.)

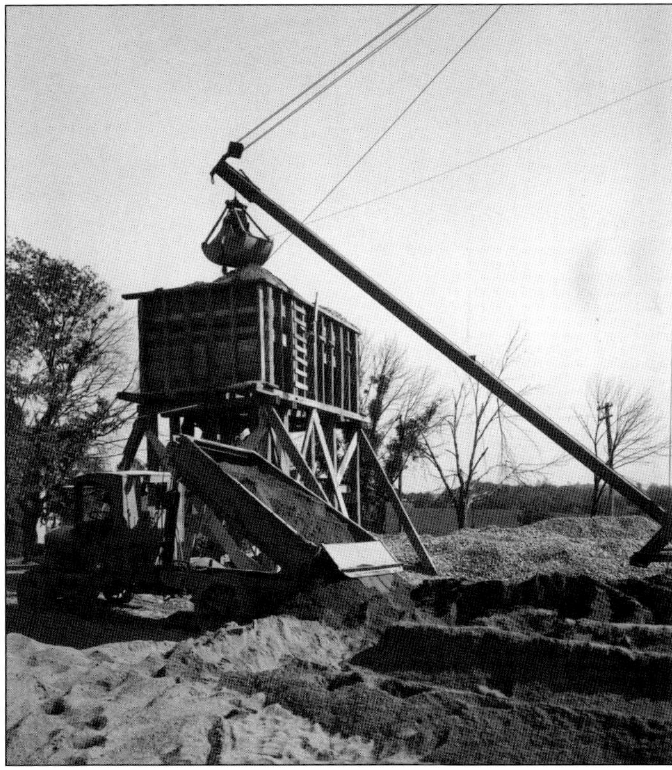

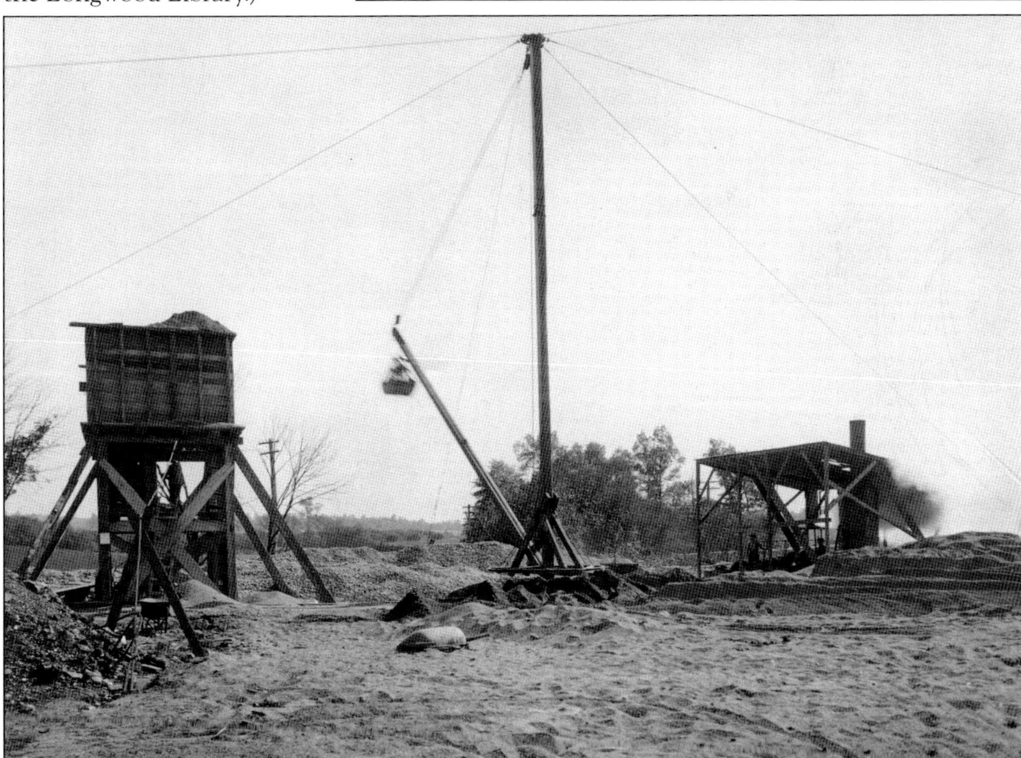

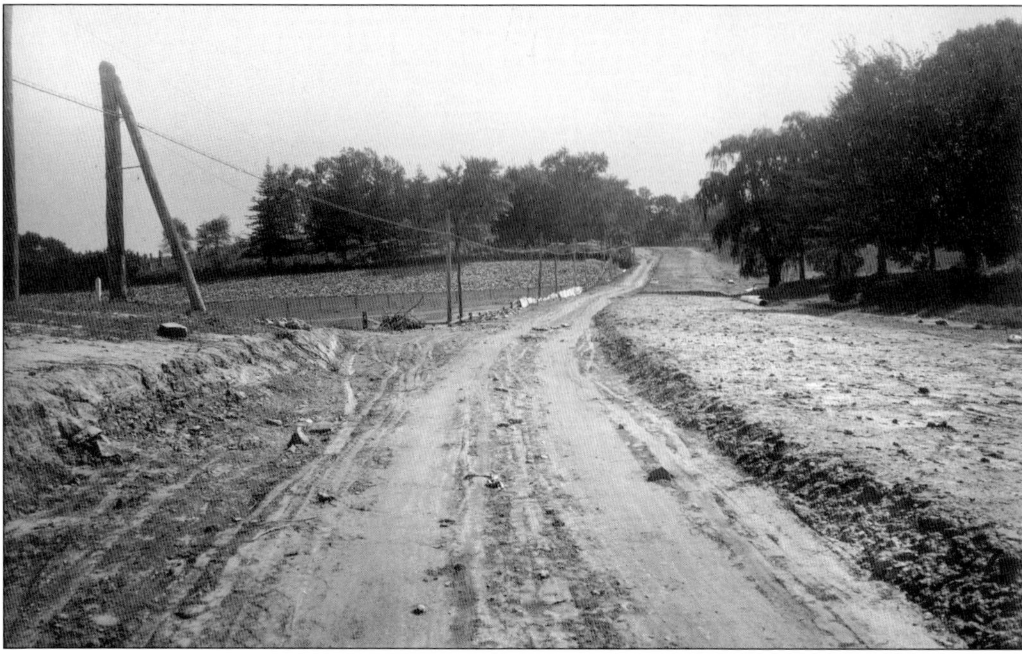

This c. 1920 view of Kennett Pike, looking from Mendenhall north toward Hamorton, shows the condition of the road before the paving and widening project undertaken by Pierre S. du Pont. This section was paved following World War I. The only completed section prior to the war was that from the state line to Mendenhall. (Courtesy of the Longwood Library.)

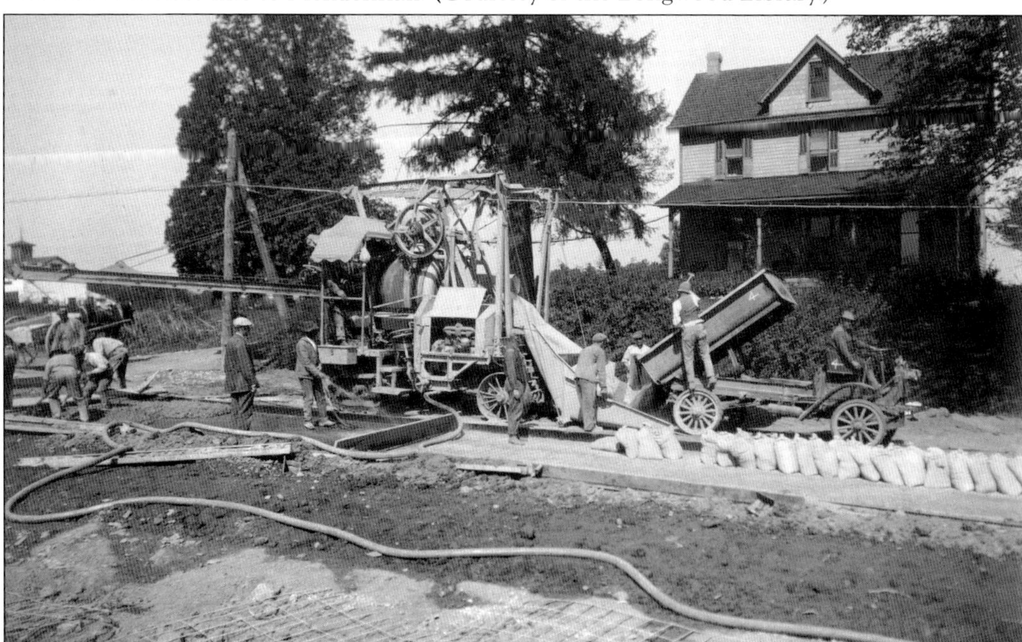

Leaving Mendenhall, travelers would next come to the village of Hamorton, Pennsylvania. This view was taken at the southern edge of the village around 1920 and shows ongoing work along Kennett Pike. Once the pike was completed in this section, the road was turned over to the State of Pennsylvania. (Courtesy of the Longwood Library.)

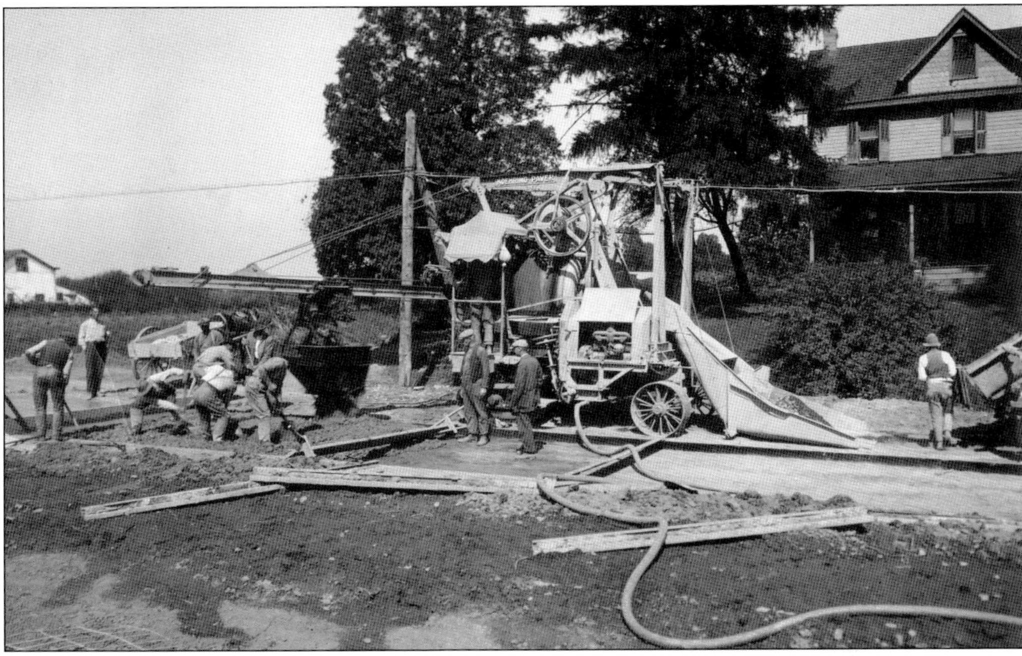

This photograph, taken in the same location as the image on the preceding page, is located near Hamorton and shows workers along Kennett Pike during construction around 1920. Subcontractors hired by the DuPont Company to work along the pike worked in crews. At its peak, nearly 140 men worked at a time along Kennett Pike. (Courtesy of the Longwood Library.)

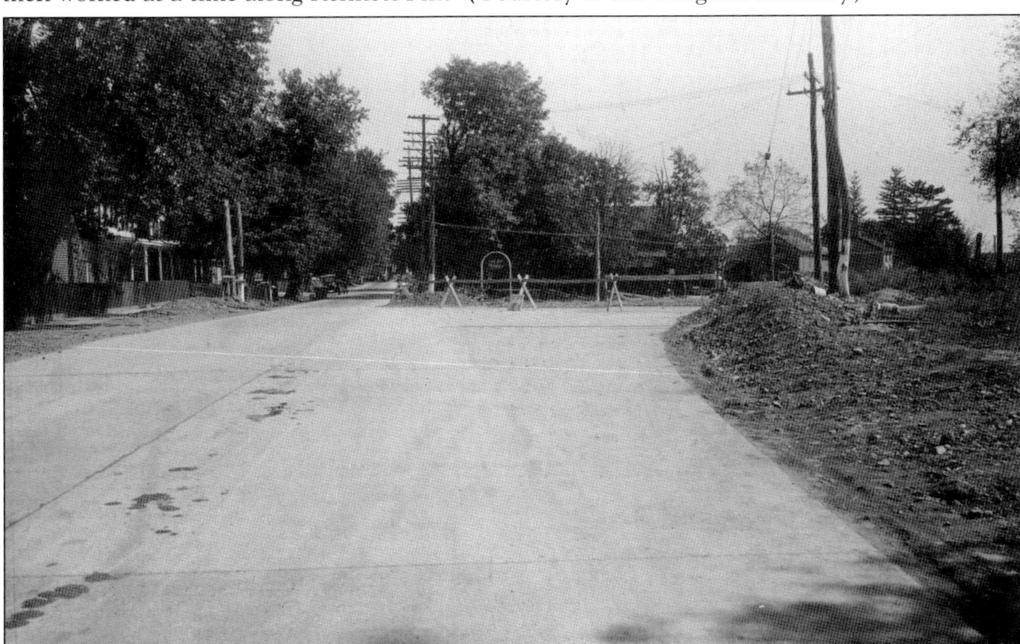

Pictured around 1920 is the intersection of Kennett Pike and Route 1 in the village of Hamorton. Kennett Pike bears right at the intersection and heads south toward Mendenhall, Centreville, and lastly, Wilmington, Delaware. The photograph shows Kennett Pike closed to traffic. (Courtesy of the Longwood Library.)

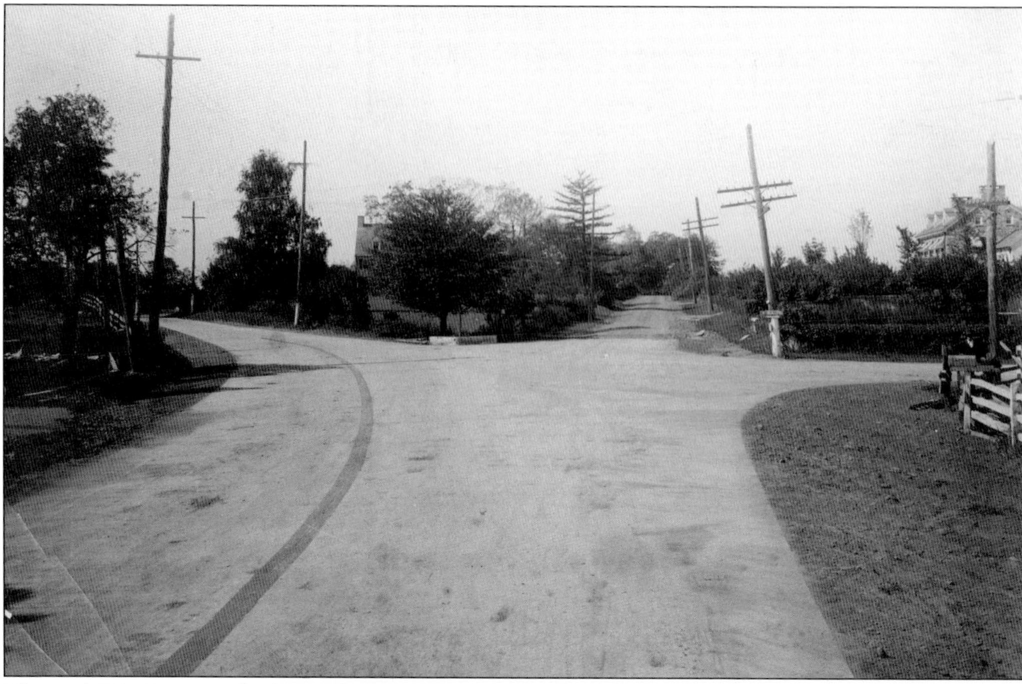

This c. 1921 photograph shows the intersection at Anvil Corner in Chester County, Pennsylvania, near Kennett Pike. In this view, US Route 1 bears to the left, Route 52 (Kennett Pike) turns right, and State Road, which led to Red Lion, is at center. (Courtesy of the Longwood Library.)

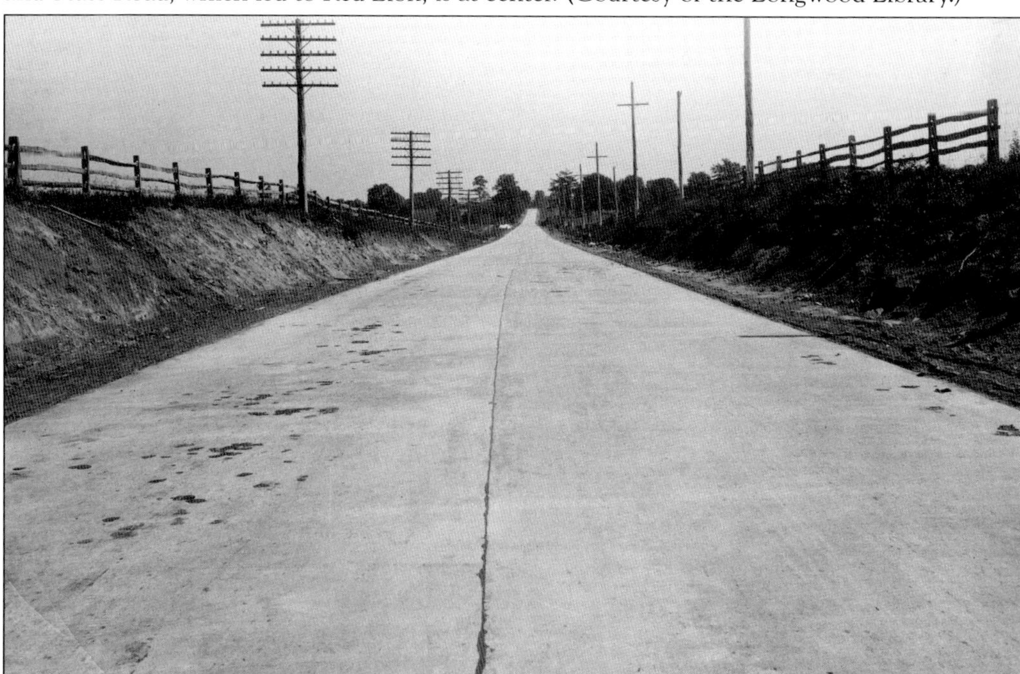

This c. 1920 photograph shows Route 1 looking east from Anvil Corner toward Hamorton, Pennsylvania. North of the main part of Route 52 on which this book focuses, this is a great view of US Route 1, which Longwood Gardens borders. (Courtesy of the Longwood Library.)

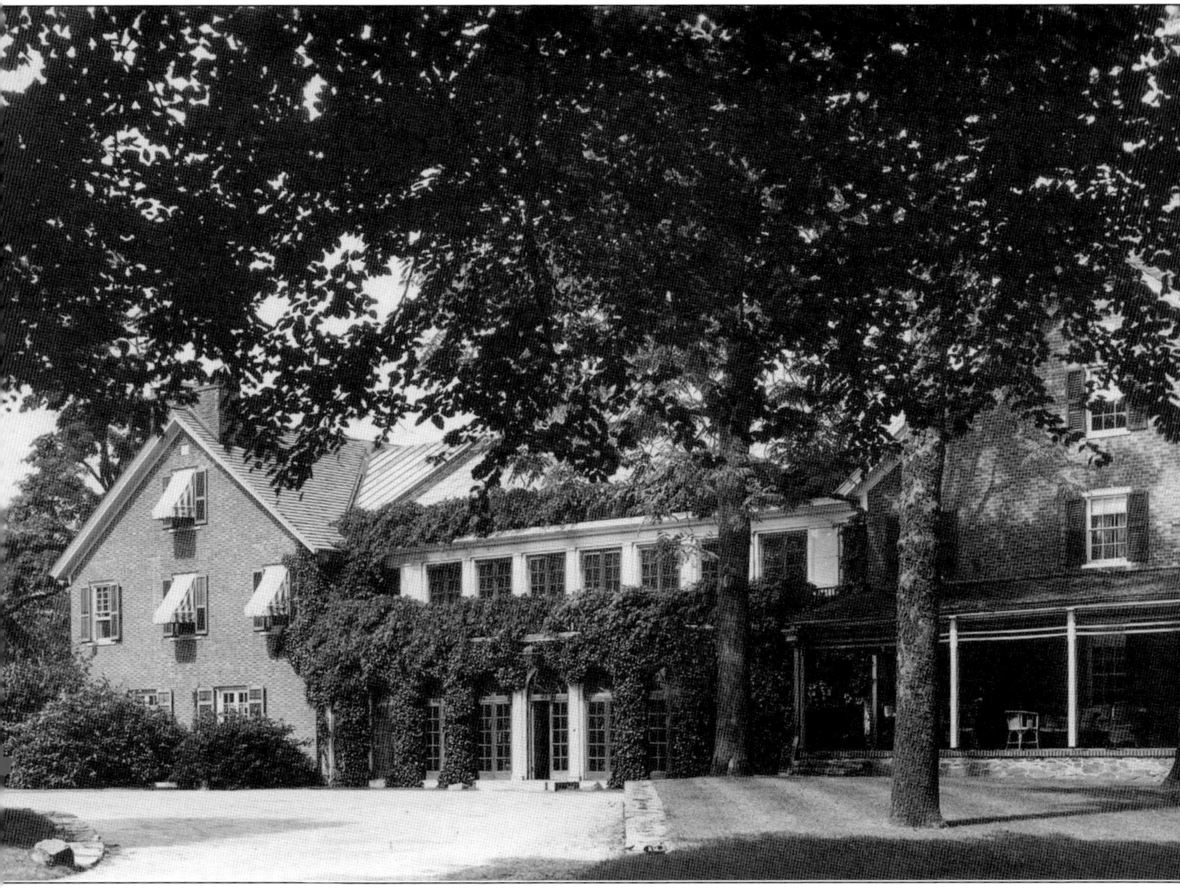

Pierre S. du Pont, financier of the Kennett Pike, made his home in Chester County just off of Route 1 near Kennett Pike at Longwood. He purchased the Peirce farm in 1906 after he learned of plans to sell what had been known as Peirce Park for timber. George Peirce purchased the original 402 acre tract from commissioners acting on behalf of William Penn. One of Peirce's sons, Joshua, built the original farmhouse on the site in 1730 (section at right). By the middle of the 19th century, the land boasted one of the country's first public arboretums. However, by 1900, Peirce family descendants lost interest in the property and had decided to sell to a lumber mill until the intervention of du Pont. After the farm and house were purchased by du Pont in 1906, two main additions (at left) were made to the original section of the house. The two sections of the house, the newer mirroring the old, were connected by a glass conservatory. (Courtesy of Hagley Museum and Library.)

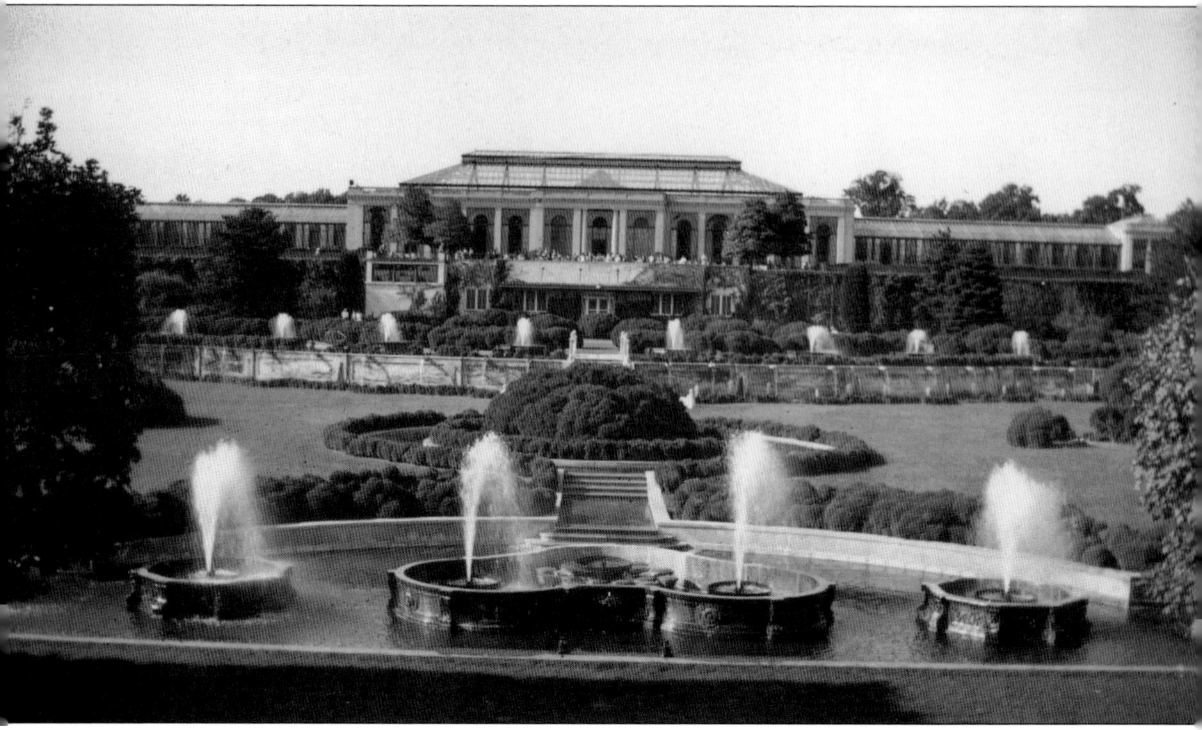

Not only did Pierre du Pont enlarge the house at Longwood, he also set out to make the garden a place to entertain friends and relatives. In doing so, he created one of the most magnificent private gardens in the country. Shown here in 1939 is the conservatory at Longwood, framed by his prized fountains. Once the conservatory was complete, du Pont worked to expand his outdoor gardens and fountain display. One of the first projects was the Italian Water Garden, which took inspiration from the Villa Gamberaia near Florence, Italy. While the Italian version lacked fountains, du Pont included more than 600 jets in his display. The Main Fountain Garden, shown in the foreground of this image, can circulate 10,000 gallons of water a minute and shoots more than 130 feet into the air. Pierre S. du Pont's home and garden are open to the public. (Courtesy of Hagley Museum and Library.)

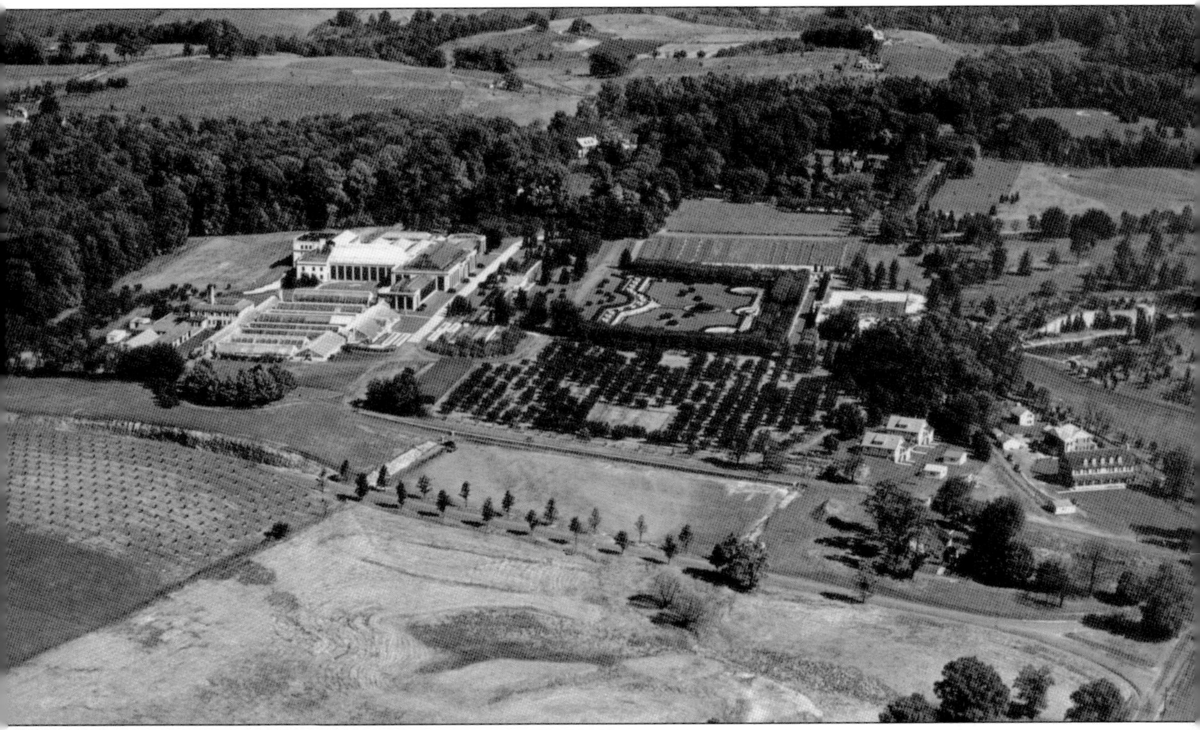

An aerial view of Longwood taken in 1934 by the Dallin Aerial Survey Company shows the enormity of the estate as well as its buildings and gardens. The main house is relatively modest for a man of du Pont's status as he focused his time and energy into creating a world-class garden. In this image, the conservatory and greenhouses can be seen at left, and the Main Fountain Garden is in the center, surrounded by open fields and tenant houses. The Peirce–du Pont House is obscured from view by the many trees that made up part of the original Peirce Arboretum. When du Pont died in 1954 at the age of 84, he had in place a well-thought-out (and -funded) plan for Longwood to continue as a public garden. (Courtesy of Hagley Museum and Library.)

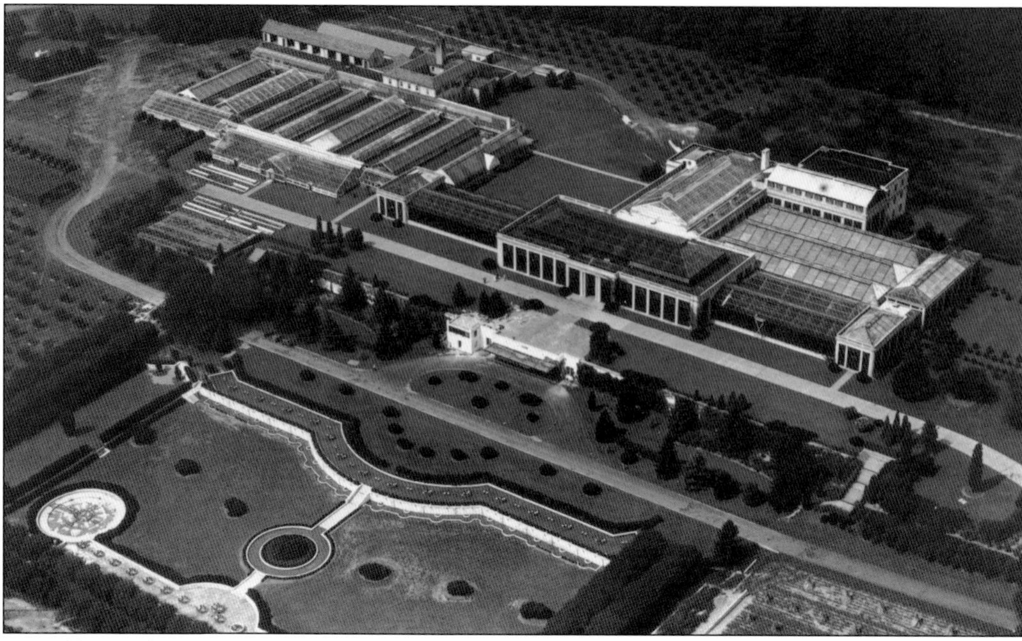

An additional aerial view of the conservatory and Main Fountain Garden in 1931, the year of its opening, shows the many greenhouses and outbuildings associated with the conservatory complex. At the rear of the building was du Pont's music room, which features a 3,650-pipe organ that still operates today. (Courtesy of Hagley Museum and Library.)

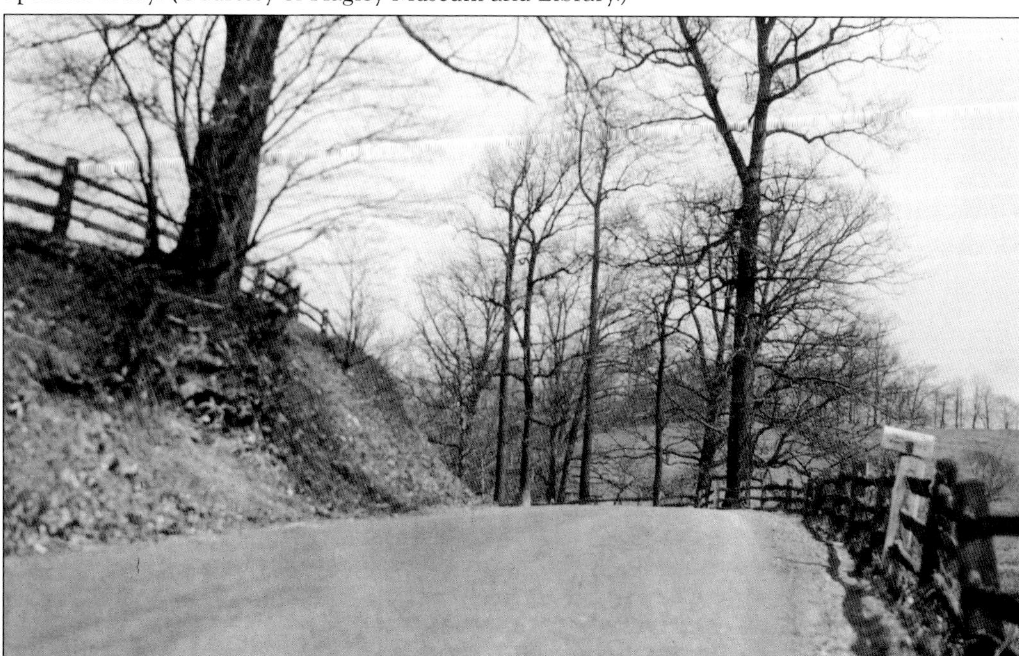

This final view of Kennett Pike in Chester County shows a curve on Route 52 between Anvil Corner and the intersection of Route 926, just northeast of Longwood. Route 52 eventually winds its way through the countryside to West Chester. Today, this image is much different; however, it is still mostly rural in nature. (Courtesy of the Longwood Library.)

Six

Kennett Pike Today

It has been almost a century since Pierre S. du Pont undertook the widening and expansion of Kennett Pike through Wilmington and into Chester County, Pennsylvania. During that time, countless changes, both good and bad, have come to Kennett Pike. In the city of Wilmington, many dwellings have been replaced with retail shops and office space. This is especially true for the area between Rodney Street and Woodlawn Avenue, where row homes have given way to Union Park's various car dealerships and convenience stores. In Greenville, where the Shields Suburban Shoppes were the only strip center in the small village, retail centers now dot Kennett Pike from Buck Road just past Hillside Road. During months when school is in session, Greenville is often jammed with traffic as Alexis I. du Pont High School welcomes students in the morning and sends them on their way in the afternoon.

Perhaps the most unchanged area along Kennett Pike in Delaware is the area around Winterthur, where the Lower Brandywine Presbyterian Church still graces the pike near the main entrance to the former du Pont estate. Following Kennett Pike north, one enters the village of Centreville. Centreville may appear to be the sleepy, quaint Delaware village it has always been, but it has also experienced its fair share of change. Many of the structures along the pike that were included in the historic district and were once home to many longtime Centreville families are now upscale retail shops. No trip to Centreville is complete without a shopping trip to Frederick's Country Corner and lunch at Buckley's Tavern.

Entering into Pennsylvania and heading toward Longwood are many of the same estates that graced the pike 80 years ago. Even those that have been divided and developed have been altered in a tasteful manner.

For all of the change that Kennett Pike has experienced, it is still one of the most beautiful highways in Delaware. In fact, in 2005, it was designated a National Scenic Byway. The beauty of Kennett Pike today is one not often surpassed. The following pages present a view of the Kennett Pike as it appears today.

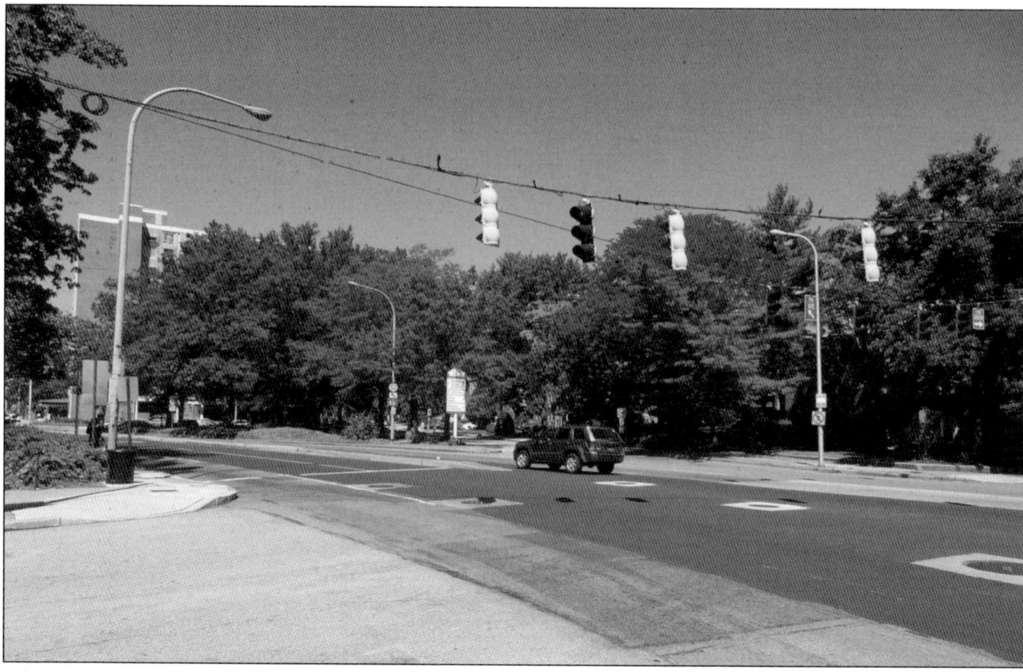

This is the view in 2013 looking north on Pennsylvania Avenue in the city of Wilmington, just north of Route 52's interchange with Interstate 95. Here, at the intersection of the avenue with North Van Buren Street, drivers can continue left to remain on Route 52 or turn to the right to travel on Delaware Avenue. With the exception of mature trees, this view has changed little in 90 years. (Photograph by author.)

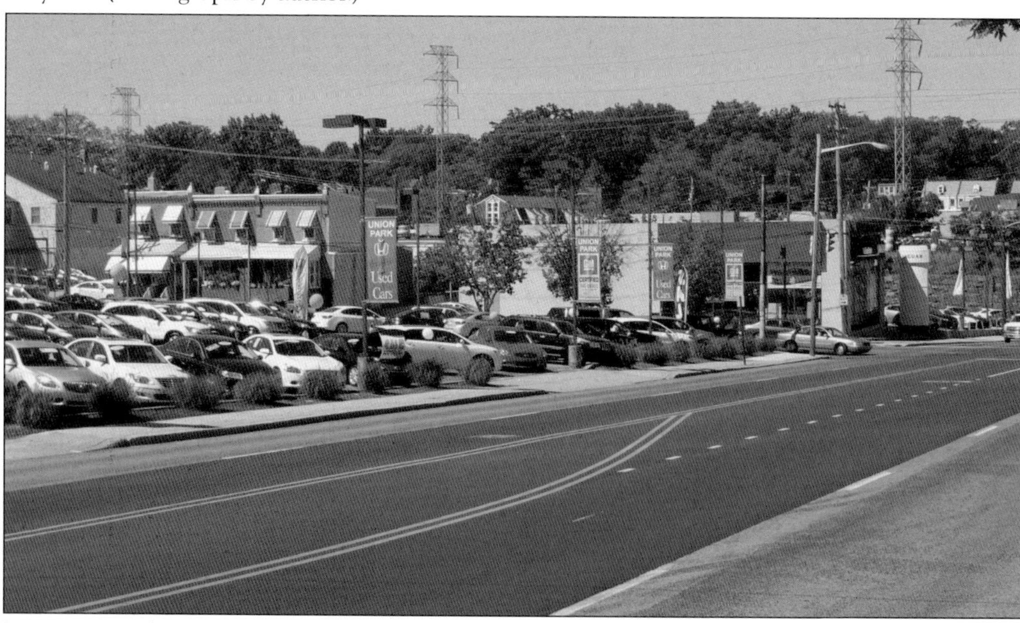

In contrast, the view along this section of Pennsylvania Avenue has changed drastically. The land where row homes and a service station once stood is now occupied by the Union Park family of car dealerships. The abutment of the train overpass can be seen at right, where the road still narrows and floods as it did in the 1920s. (Photograph by author.)

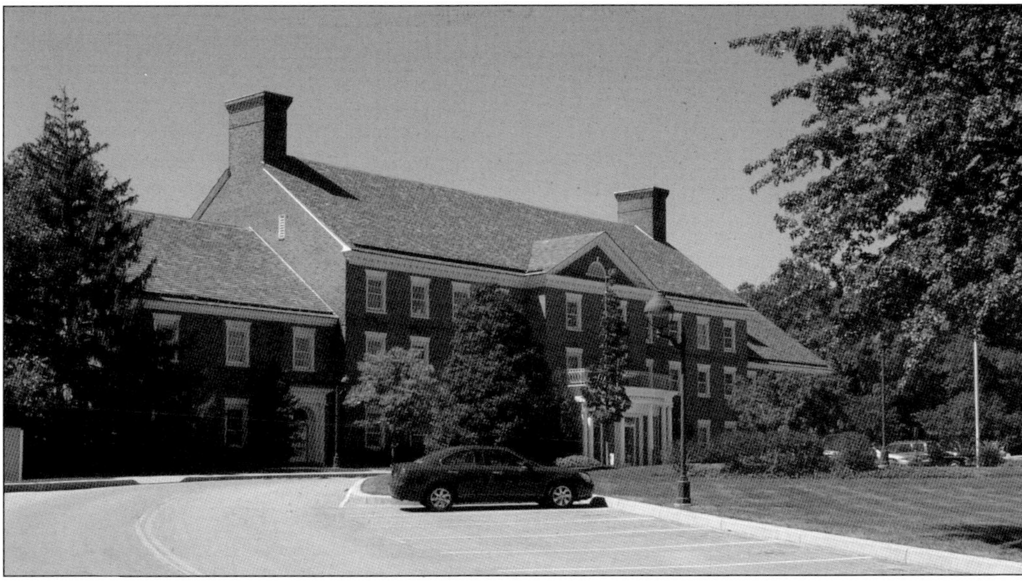

Shown above is the Eugene du Pont Preventative Medicine & Rehabilitation Institute of Christiana Care, located at Pelleport. After the du Pont family mansion was demolished in 1954, Eugene du Pont donated $1 million as an endowment to build a hospital on the site. Originally named the Eugene du Pont Convalescent Memorial Hospital, the 60-bed hospital opened in 1955. Pelleport is located at the intersection of Montchanin Road (Route 100) and Kennett Pike (shown below). Properties in the vicinity have largely remained the same. Farther down Montchanin Road (center of the image), the Raskob Foundation for Catholic Activities is located at right, and St. Joseph on the Brandywine Church, at left. (Both, photograph by author.)

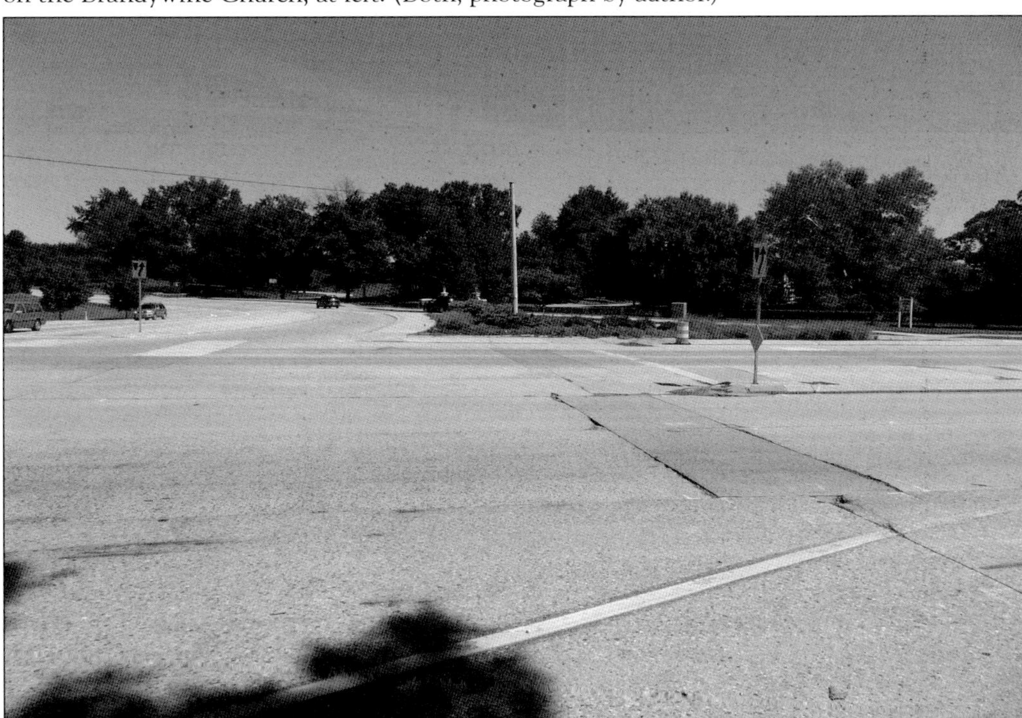

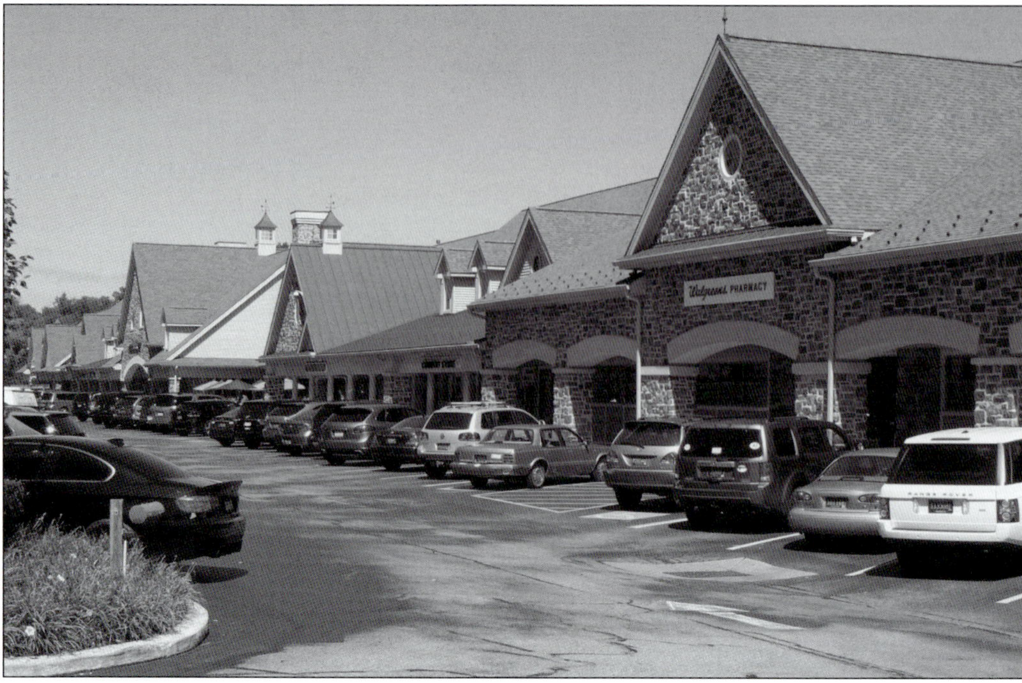

Pictured here are two views of Greenville, Delaware. The image above shows what used to be the Shields Suburban Shoppes located along Kennett Pike. All but one (Wilmington Country Store) of the original occupants of the shopping center have either closed or relocated elsewhere in the area. The shopping center was recently renovated and has for a number of years been called One Greenville Crossing. The image below is a view of Kennett Pike in Greenville looking north toward Centreville. A similar view from the early 20th century appears earlier in this book and shows a more remote and rural Greenville. (Both, photograph by author.)

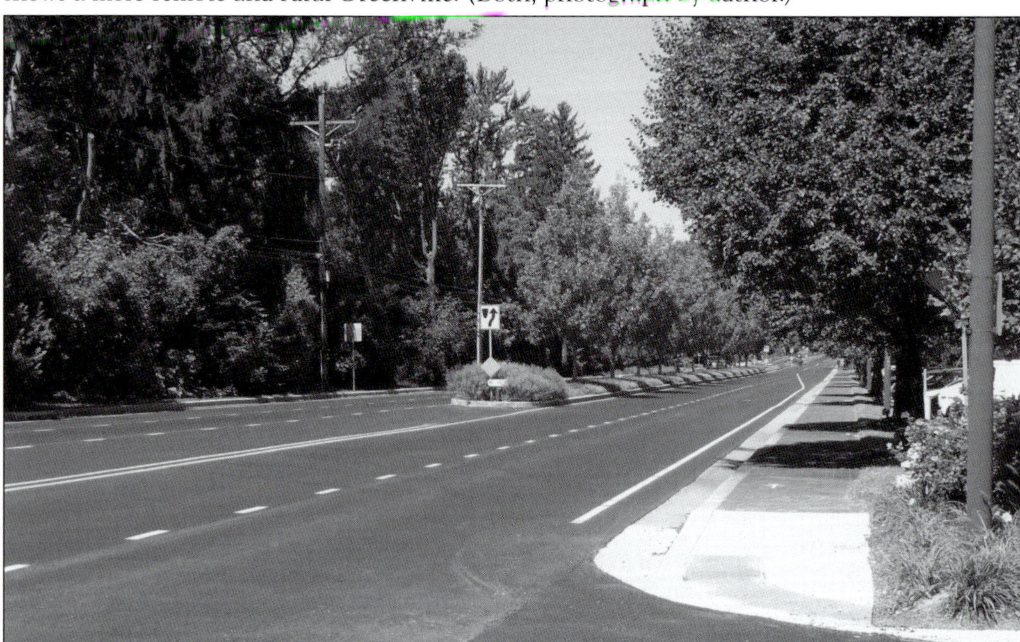

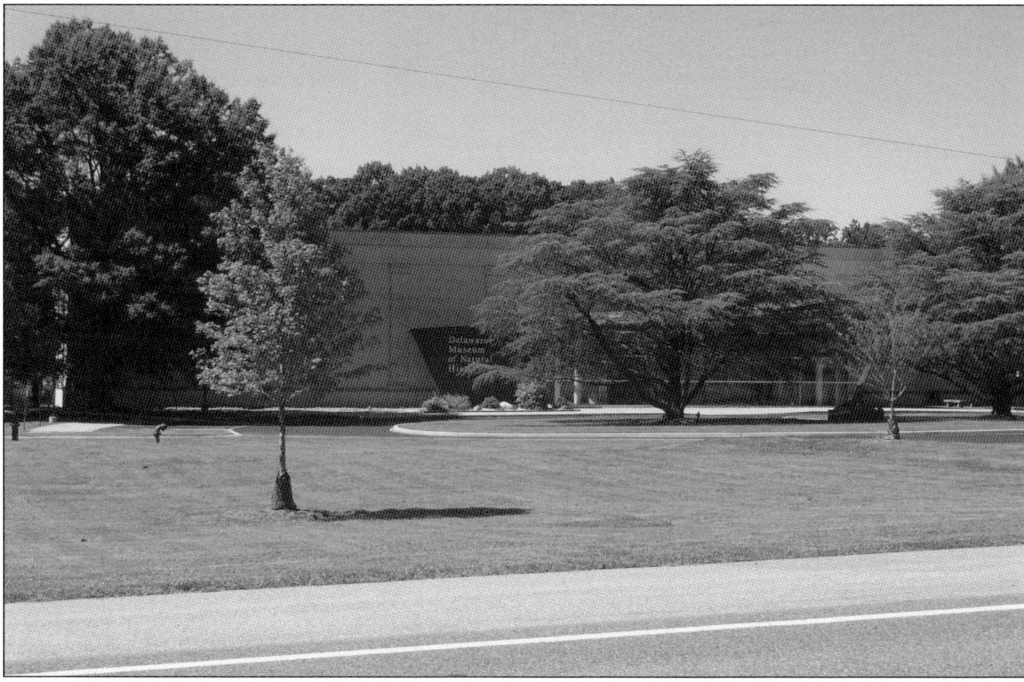

The Delaware Museum of Natural History is located along Kennett Pike's western edge, near Henry Francis du Pont's Winterthur Museum. The museum, founded by John E. du Pont, was opened to the public in 1972 and featured his personal collection of mollusks and birds. The land on which the facility is located was given to John du Pont by his cousin Henry Francis du Pont. (Photograph by author.)

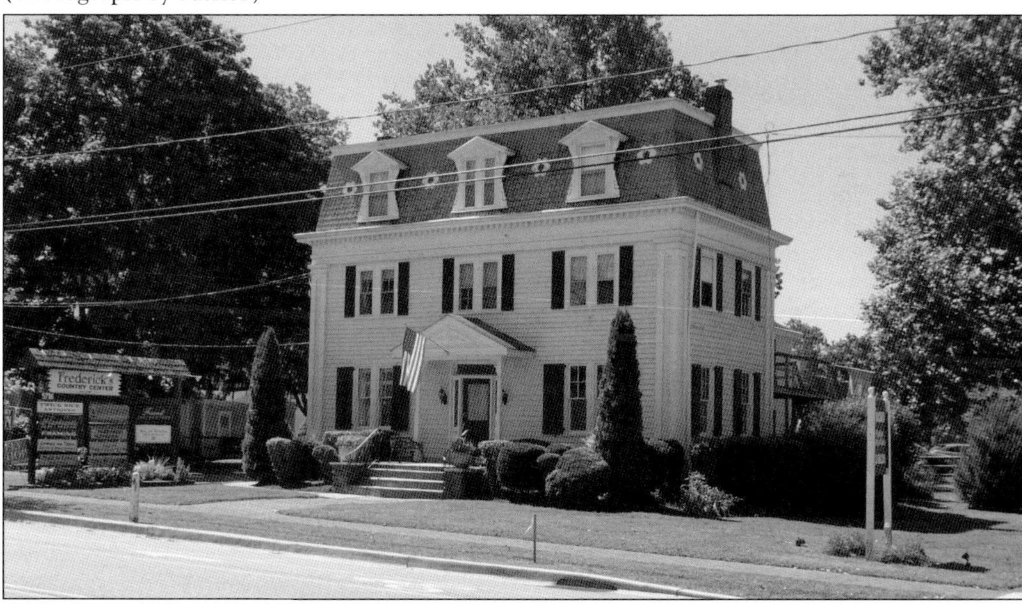

Frederick's Country Center is perhaps the most popular shopping location within the village of Centreville, Delaware. The Frederick family moved into the house more than a century ago, and in the 1950s, Edward P. Frederick began converting the outbuildings behind the house into retail shops that today include a wide variety of stores. (Photograph by author.)

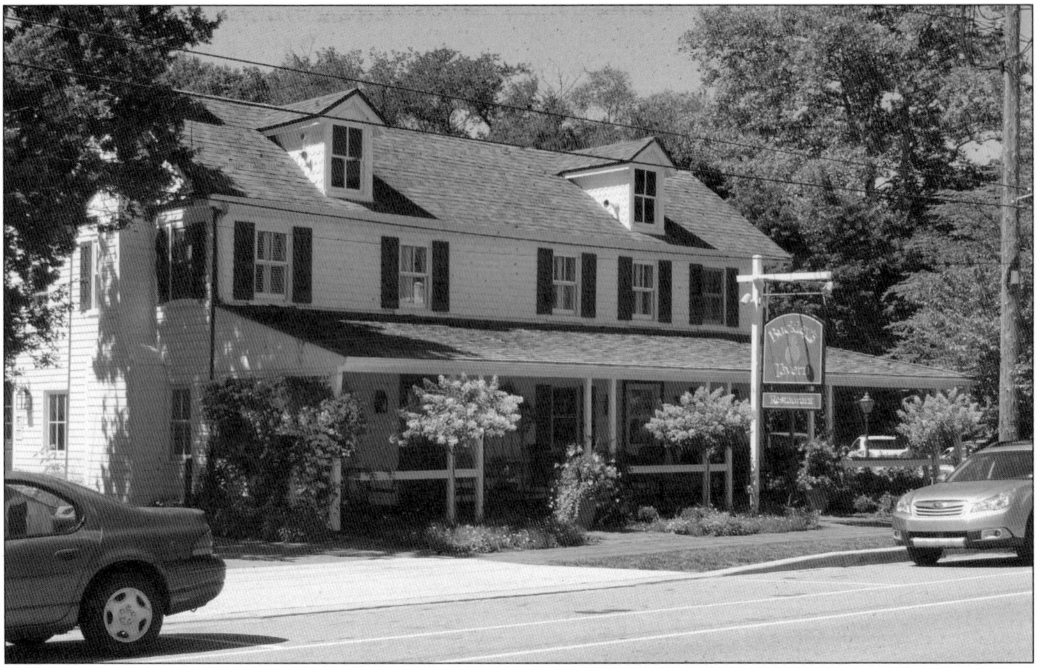

The popular Buckley's Tavern is pictured here. Located in Centreville, it was recently closed and renovated and is now open under new management. The restaurant building dates to the mid-19th century. The parking lot at Buckley's was the site of Ted Antoine's service station from 1926 into the late 1940s. (Photograph by author.)

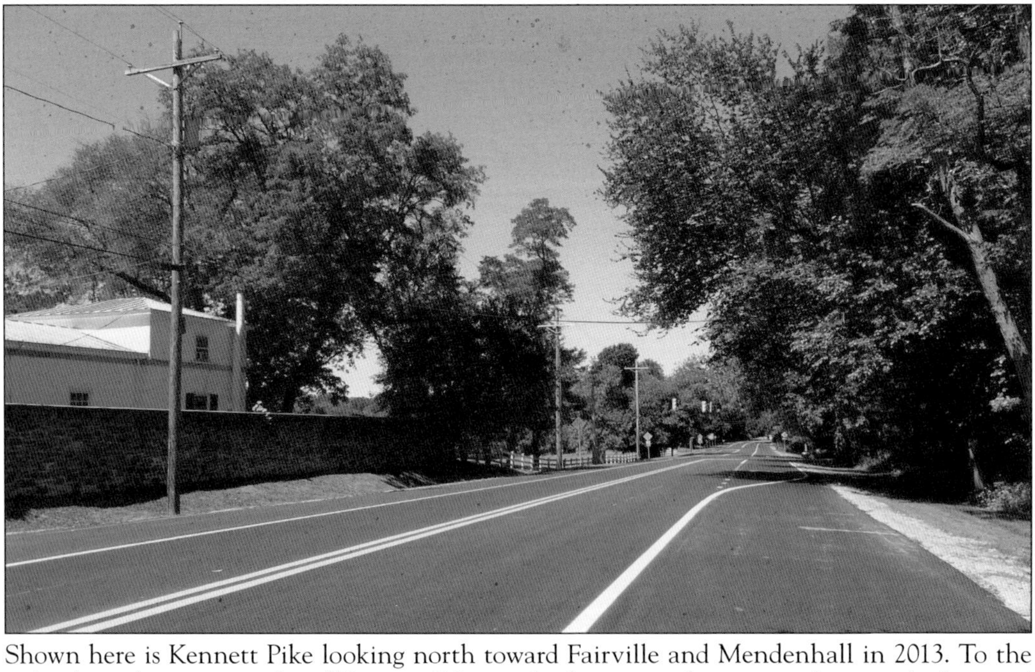

Shown here is Kennett Pike looking north toward Fairville and Mendenhall in 2013. To the right along the western edge of Kennett Pike are the former gatehouses to what was the Lungers' Oberod. One of the homes was built in the 1850s and was the home of Sen. Daniel Hastings. The other dwelling on the site is a 20th-century addition. (Photograph by author.)

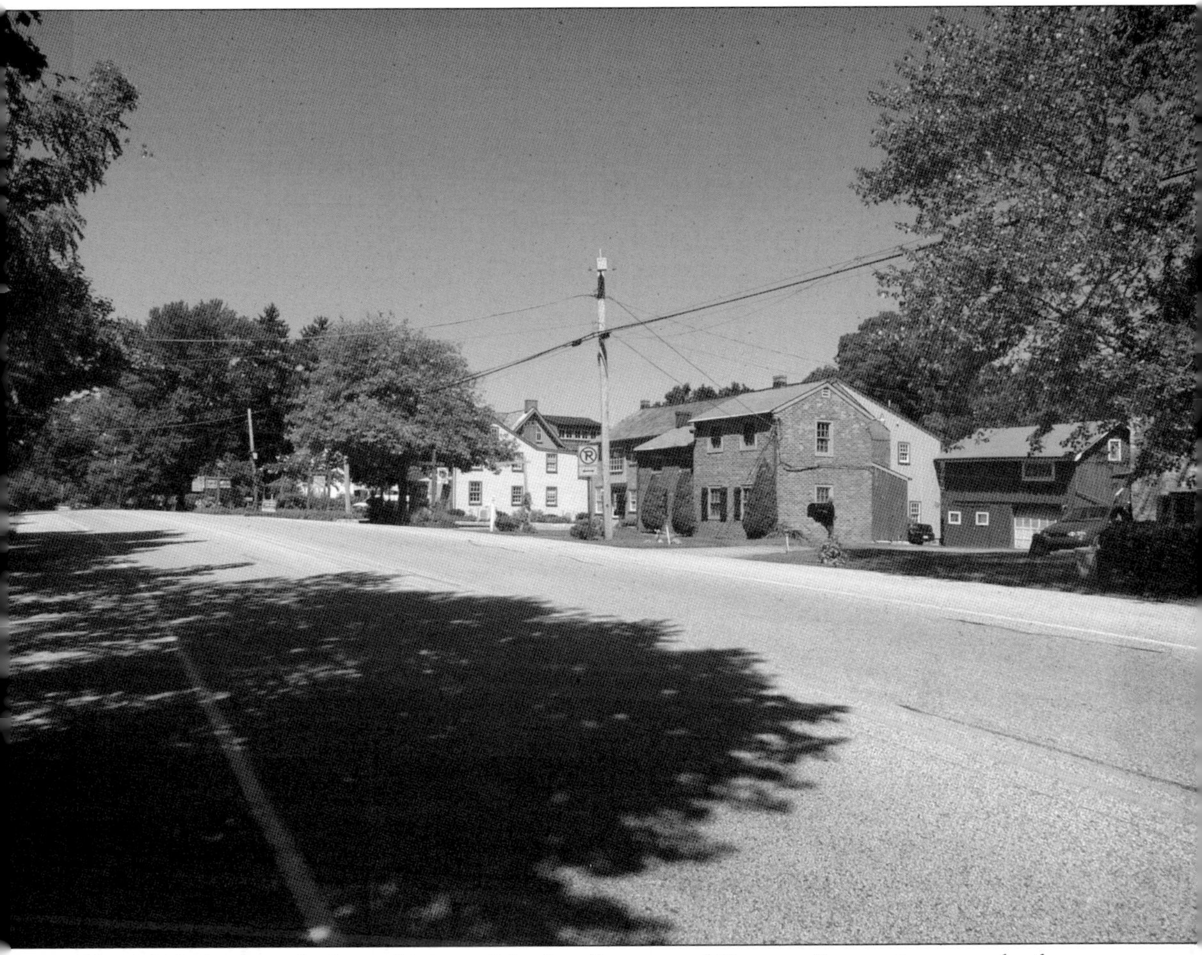

This 2013 photograph shows Kennett Pike heading toward Kennett Square just past the former estate of Philip du Pont. Fairville was developed in the early part of the 19th century. The village proper lies along the eastern side of the pike. The businesses that were located along the stretch of road served local residents and farms. Located in the village just off of Kennett Pike is the Harlan Log House, which was built around 1715 and is one of the best-preserved log structures in Chester County. Today, Fairville's buildings have been adapted for reuse and are used as offices and galleries. One of the most popular destinations in the village is the Fairville Inn, built sometime between 1825 and 1837. The original summer kitchen and root cellar still remain intact on the property. The inn features a total of 15 rooms in three separate buildings and has been featured in the *New York Times* as a travel destination. (Photograph by author.)

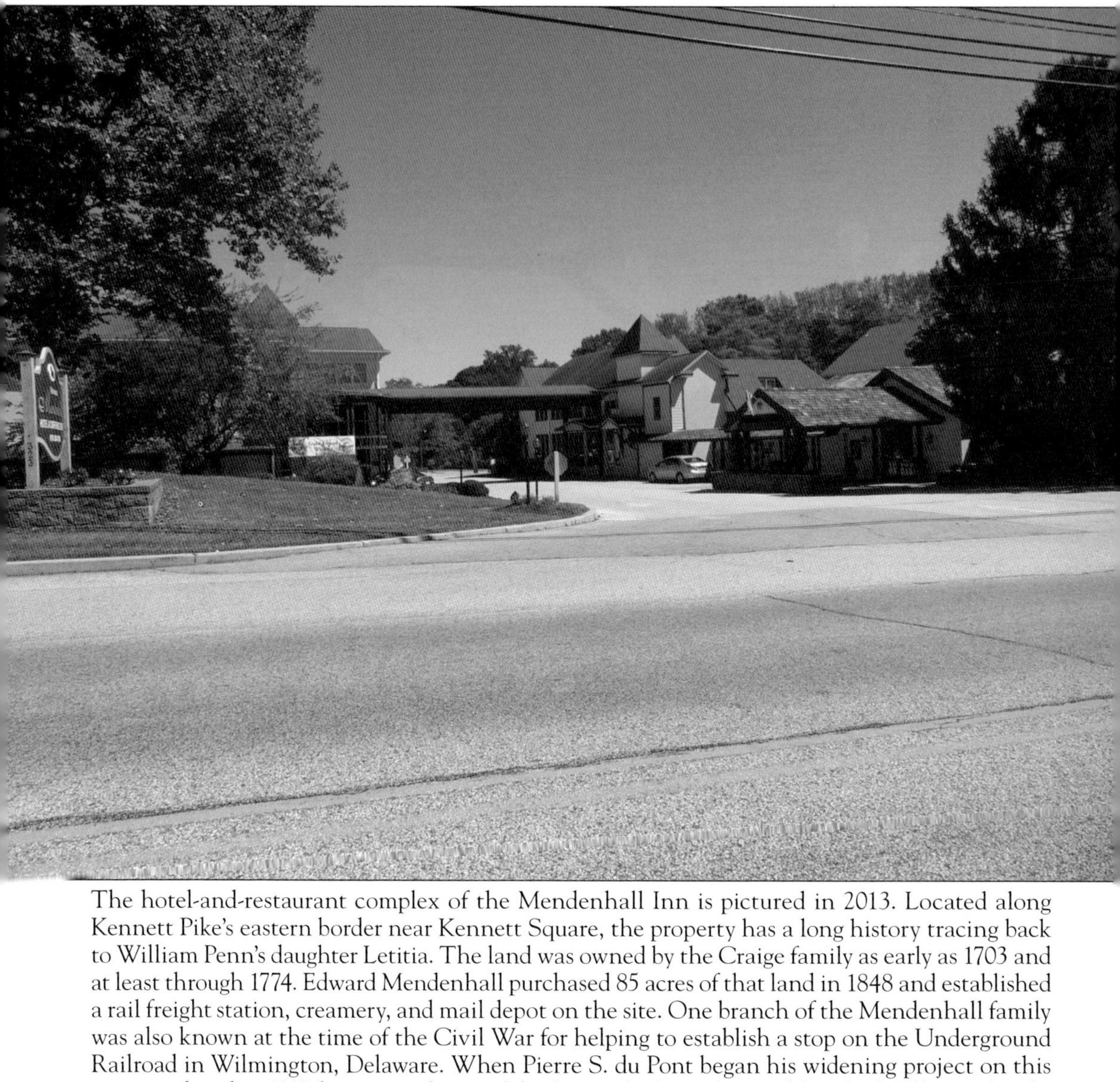

The hotel-and-restaurant complex of the Mendenhall Inn is pictured in 2013. Located along Kennett Pike's eastern border near Kennett Square, the property has a long history tracing back to William Penn's daughter Letitia. The land was owned by the Craige family as early as 1703 and at least through 1774. Edward Mendenhall purchased 85 acres of that land in 1848 and established a rail freight station, creamery, and mail depot on the site. One branch of the Mendenhall family was also known at the time of the Civil War for helping to establish a stop on the Underground Railroad in Wilmington, Delaware. When Pierre S. du Pont began his widening project on this section of road in 1917, he acquired part of the farm, which was later sold to the Millhouse family. The Millhouses converted the barn on the property into a feed store. It was later owned by the Pusey family and then by Hitch's Lumberyard. The barn was turned into a restaurant in 1968 and was bought by the Hionis family, who added the hotel and grand ballroom. (Photograph by author.)

Bibliography

Cannon, Patt. *Centreville: The History of a Delaware Village, 1680–2000*. Centreville, DE: Centreville Civic Association, 2001.

Janson, Donald. "Du Pont's Toast a New Museum." *The New York Times*. May 13, 1972.

Laird, W.W. *For It's Always Fair Weather*. Cedar Tree, 1979.

Lidz, Maggie. *The Du Ponts Houses and Gardens in the Brandywine*. New York: Acanthus, 2009.

longwoodgardens.org/history

"National Register of Historic Places—Centreville, Delaware." *National Register Nomination*. National Park Service, 06 June 2013. http://pdfhost.focus.nps.gov/docs/NRHP/Text/64000101.pdf.

"National Register of Historic Places—Rodney Court." *National Register Nomination*. National Parks Service, 1980. http://pdfhost.focus.nps.gov/docs/NRHP/Text/80000938.pdf.

Riggs, John B. *A Guide to Manuscripts in the Eleutherian Mills Historical Library*. Greenville: Eleutherian Mills Historical Library, 1970.

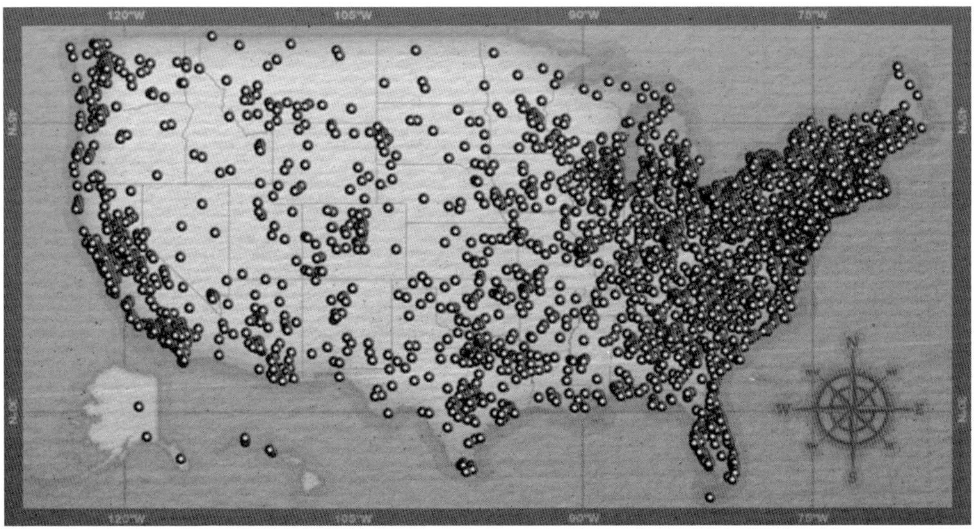